No. 1193
$22.95

Broadcasting Around The World

by William E. McCavitt

TAB BOOKS Inc.
BLUE RIDGE SUMMIT, PA. 17214

FIRST PRINTING

FIRST EDITION

Copyright © 1981 by TAB BOOKS Inc.

Printed in the United States of America

Reproduction or publication of the content in any manner, without express permission of the publisher, is prohibited. No liability is assumed with respect to the use of the information herein.

Library of Congress Cataloging in Publication Data

McCavitt, William E

    Broadcasting around the world.

    Includes index.
    1. Broadcasting. I. Title.
HE8689.4.M32   384.54   80-19929
ISBN 0-8306-9913-9

Contents

 # Preface

This book came about because of the interest developed at Syracuse University in the early '60s in the field of international broadcasting. It was during this time, as a graduate assistant, that I came in contact with broadcasters from various countries around the world as they came to this country to study our system of broadcasting. Over the years I have had an opportunity to visit some of these countries and study their particular approach to broadcasting.

We can always learn from others and this is true in broadcasting also. This book will give students of broadcasting a chance to read about radio and television in other countries. Each chapter has been written either by an individual involved in broadcasting in their respective country or has been contributed by the official broadcast system of the country.

Compiling such a book has presented some unique experiences such as calling Australia and being put on hold, hearing our campus operator saying "You have a call from Jerusalem," and having a chapter arrive with one page missing. The cooperation of the contributors has been excellent and I appreciate the time and effort put into each chapter by the authors and organizations involved:

South Africa: Dr. T. L. de Koning, Rand Afrikaans University
Poland: Professor Stanislaw Miszczak, Warsaw University
Union of Soviet Socialist Republics: Dr. Valdimir Yaroshenko, Moscow State University
Japan: Public Relations Bureau, Japan Broadcasting Corporation
Korea: "Annals of Newspapers & Broadcasting in Korea 1979," published by Korea Research Institute of Journalism
Israel: Mr. Ari Avnerre, counselor to the director general, Israel Broadcasting Authority
Canada: Mr. Albert A. Shea, senior planning officer, Canadian Radio-Television & Telecommunications Commission
United States: Dr. Bob Wallace, Edinboro State College, Edinboro, PA

Brazil: Mrs. Esmiralda Teixeira, Departmento Nacional De Telecomunicacoes

Guyana: Mr. Ron Sanders, communications consultant to the Government of Antigua

India: Ms. Mehra Masani, vice-president, International Institute of Communications

Great Britain: British Information Services, Central Office of Information

Germany: Dr. Richard Dill, coordinator of international relations, Rundfunkanstalten der Bundesrepublik Deutschland

Ireland: Louis McRedmond, head of Information & Publications, Radio Telefis Eireann

Italy: Italian Radio Television

Netherlands: Dr. Hans Van Den Heuvel, Government Department of Culture, Recreation & Social Welfare

Sweden: Mr. Herbert Soderstrom, information director, Swedish Broadcasting Corporation

Australia: Mr. Peter Lucas, senior officer, Australian Broadcasting Commission

Special thanks must go to TAB Books for their willingness to publish such a book, the International Institute of Communication for their cooperation and encouragement, and last but not least to Helen Clinton, my graduate assistant, who helped keep the project moving.

William E. McCavitt

Chapter 1
Broadcasting in the
Republic of South Africa

Prof. T.L. de Koning
Rand Afrikaans University

Not only political factors, but also geographic, economic, and cultural factors determined the existing broadcasting system in the Republic of South Africa (RSA). Broadcasting in the Republic of South Africa is run and controlled by the giant corporation, the South African Broadcasting Corporation (SABC). In principle the SABC is a state-subsized monopoly and, apart from the board, appointed by the minister of national education, the SABC is supposedly autonomous. Throughout the history of the SABC, however, both the Nationalist and United Party governments have been accused of nepotism in key appointments and of misusing the broadcasting medium for political aims.

THE SHAPING OF SOUTH AFRICAN BROADCASTING

To fully appreciate the present system of broadcasting in the RSA, it is necessary to look at those factors which determined the present system of broadcasting as well as the historical development of the SABC.

Geographical Factors

The RSA stretches over an area of 1,178,679 km, between 22 degrees south to 35 degrees south latitude and 17 degrees east to 33 degrees east longitude. This area includes every possible climatic region from subtropical through semi-desert to Mediterranean.

The topography of the RSA can roughly be compared to an inverted saucer which rises in two steps to the Highveld. Since the country is triangular-shaped, a long coastline forms the eastern, southern and western borders. A very narrow coastal plain of 5 to 15 miles is bordered by the first mountain barrier of 4000 to 5000 feet. Beyond this mountain barrier lies the Small Karoo, a region of sloping hills and narrow valleys, changing into the plateau region of the Great Karoo which is bordered by the brim of the second step. This brim of the escarpment reaches heights of 3299 meters in the Drakensberg mountains. North and west of the Drakensberg mountains is the Highveld of Southern Transvaal, Free State and Northern Cape. The Transvaal Highveld is encircled in the west, north and east by the Soutpansberg, Waterberg, Megaliesberg, Lebombo and Maluti mountain ranges.

The vast distances and mountain barriers demanded a network of powerful transmitters in order to cover the whole area of the republic. Apart from the concentric mountain ranges, lightning and sunspots pose electrical interference which present radio engineers with severe problems. Supplying the power necessary to overcome the above obstacles with AM transmissions limited the SABC to those urban areas which could support their local stations.

Language and Cultural Differences

Broadcasting, as is politics, is shaped by the diversity of population and cultural groups. The population, excluding the homelands, consists of:

Blacks	16,279,000
Whites	4,310,000
Coloreds	2,426,000
Asians	746,000

The Republic of SA Yearbook, 1977 (1976 figures), p. 75

The black population group does not constitute a homogeneous entity. Each group within the black population has its own language and cultural traditions. It should be emphasized that there is no such thing as a common black language or a common black culture. The following language and ethnic groups exist in the Republic, excluding the Transkei:

North Sotho	2,011,000
South Sotho	2,627,000
Zulu	5,004,000
Tswana	2,103,000
Xhosa	872,000
Sjangaan	814,000
Swazi	590,000
Ndebele	615,000
Other	2,194,000

The above from The Republic of SA Yearbook, 1977 (1976 figures), p. 75.

Apart from the Black diversity of culture and language, the white population consists of Afrikaans and English sections. Therefore, the RSA has two official languages, Afrikaans and English, which by law must receive equal priority.

A breakdown (SA Yearbook, 1977) of the Asian, white and colored population according to language appears in Table 1-1.

The SABC has always taken positive steps to deal fairly with the differences posed by the existence of numerous languages, together with profound cultural differences. The limited frequencies available on AM, and distortion of the signal due to topographical factors, posed economic and technical difficulties which were insurmountable before the introduction of a comprehensive FM network in the 1960s.

The vast area to be covered and the sparsely populated rural areas prohibited a truly national broadcasting system. The republic as a whole has a density of only 18 per km. The depression of 1933 and the economic upswing and industrialization since 1940, following World War II, have led to a continuing urbanization. However, this urbanization is concentrated in a few major cities, especially the area of the Rand. These urban centers are Pretoria and Johannesburg in the north, Cape Town in the west, and Durban in the east, with Port Elizabeth halfway between the ports of Durban and Cape Town and Bloemfontein between Cape Town in the south and Johannesburg in the north.

The percentage urbanization of the blacks (33%) would have been much higher if it were not for influx control. Nevertheless, a black service and even a national service for whites, covering the whole area of South Africa, would not have been an economically feasible proposition due to the tremendous cost involved to set up the facilities necessary for this coverage.

Another point which should be kept in mind is that industrialization began in the RSA after World War II. Apart from the fact that the diverse and unequally developed populace posed problems in terms of programming, a large proportion of the population, especially the blacks, could not

Table 1-1. Breakdown of South Africa's Population by Ethnic Origin.

Home language	Asians	Whites	Coloreds
Afrikaans	0.7%	56.9%	78.9%
English	28.5	37.2	6.0
Afrikaans and English	5.3	1.0	14.8
German	-	1.4	-
Dutch	-	0.6	-
Tamil	24.4	-	-
Hindi	18.5	-	-
Gudjurata	7.3	-	-
Urdu	6.2	-	-
Percentage Urban Population			
White	Asian	Colored	Black
86.8%	74.1	86.7	33.0

afford a radio receiver. The small number of the better-off economic white population did not justify an electronic industry. With the advent of the cheaper transistor radio the possibility of owning a radio was within reach of everybody. The possibility of a viable electronics industry is also the major reason why the RSA introduced television relatively late.

The problems of RSA broadcasting is summed up by Orlik as follows:
"It is as though the United States tried in one stroke to serve all the states East of the Mississippi from Washington DC without any local stations being in existence. The massive construction program that would be required to implement such a project would severely tax even U.S. entrepreneurs, and in South Africa there would be no commercial funds to amortize the scheme."

It must be remembered that advertising revenue could not become a source of income before industrialization and before the widespread ownership of radios (Orlik, P., p 5). The only income available was from listener licenses. The SABC is a government monopoly, but it is doubtful whether private enterprise could develop broadcasting in the RSA in the face of such tremendous odds. Today, because of political problems, the RSA is relatively isolated from the West.

As the most advanced nation in Africa and with an expanding economy, the RSA is looking toward the awakening of rest of Africa for economic and political ties. South Africa currently generates 60% of all the electricity used in Africa and, therefore, has the electrical capacity necessary to conduct large-scale broadcasting operations. The positive economic situation has benefited a growing electronics industry to such an extent that with the introduction of television all the television sets could be manufactured locally. Today the SABC has an overall budget of R133 million with an income from license fees of R56 million and advertising revenue amounting to R62 million (SABC Annual Report, 1978, p 29).

As broadcasting monopoly within a diversified and heterogeneous population, it is the more essential that the SABC should be above politics and appointments of key personnel without nepotism. Unfortunately, this is not always the case.

HISTORICAL PERSEPCTIVE

On July 1st, 1924, the Associated and Technical Societies of Johannesburg started the first regular radio transmission in South Africa. In the same year the publicity associations of Cape Town and Durban followed suit.

In 1927, the three independent studios found themselves in financial difficulties mainly due to the small number ($\pm$ 5000) of licensed listeners. The Schlesinger organization came to the rescue by taking over the Johannesburg, Cape Town, and Durban operations to form the African Broadcasting Corporation.

Development of The SABC

The African Broadcasting Corporation became a financial success as licensed radio listeners steadily increased to 73,704 (Union Office of

Census and Statistics, Official Year Book, 1933/1934, p. 750). However, the very fact that broadcasting had become so integral a part of citizens' lives, the government was compelled to assume responsibility for a national policy on broadcasting.

The then Prime Minister Hertzog invited BBC Director-general Sir John Reith to South Africa during 1934, with a view on formulating policy for the future development of radio. Sir John Reith's report on broadcasting policy and development, released in March, 1935, advocated a state-controlled system in order to provide national coverage, since due to technical difficulties no commercial company could be expected to provide truly comprehensive programs nationwide. The "difficulties" referred to by Sir John Reith included geographic, economic, sonal, cultural and political factors which until today are closely related to the broadcasting scene in South Africa.

On the strength of Sir John Reith's report, Act No. 22 of 1936 was adopted in parliament and the African Broadcasting Company (whose 10-year license had expired) was officially dissolved in favor of the new South African Broadcasting Corporation. Many of the conditions and ideals of the BBC Charter were incorporated into the South African Broadcast Act, which provided for a nine-member board of governors appointed by the governor-general-in-council, as well as advisory councils at Johannesburg, Pretoria, Cape Town, Durban, Pietermaritzburg, and Grahamstown.

From the beginning the SABC was, apart from technical problems, confronted with the problem of compiling and presenting its broadcasting with due regard to the interests of both the English and Afrikaans cultures. In 1938, the then board of governors of the SABC stated that, apart from the fact that the members of the board were appointed by the governor of South Africa, the SABC was in terms of programs and internal policy an autonomous body independent of the government. However, the SABC was subject to the regulations imposed by the minister of post and telegraphs to whom annual reports should be submitted. At present, the position in terms of autonomy is practically the same as in 1936, except that the minister of post and telegraphs has been replaced by the minister of national education as the responsible minister. Advisory councils for the main cities have been replaced by advisory committees for: services in Bantu languages, Afrikaans language, English language, English religious services, Afrikaans religious services, and an advisory committee for coloreds on radio.

The Nationalist Party came to power in 1948. The present statutory structure of the SABC was, therefore, inherited from the United Party and not, as alleged, a tool or structure developed by the Nationalists to strengthen the policy of apartheid. Under the United Party government the Nationalists complained vehemently about the subordinate role of Afrikaans, the fact that Afrikaans transmitters were much weaker (0.06 kW to 2.0 kW) compared to English (10 kW), lack of an Afrikaans transmitter e.g. Durban, and blatant political broadcasts by the then Prime Minister General Smuts. Nationalist Member F. C. Erasmus suggested in parliament on September 21, 1938, that the best policy was to keep broadcasting completely out of politics. Nationalist Party complaints increased with the wartime broadcasts of General Smuts.

Within the ranks of the SABC, dissatisfaction and insubordination as a result of the United Party political broadcasts came to a head with the reorganization of the pro-government staff and the requirement to sign a loyalty oath to the SABC and its avowed object of fully cooperating with the government.

The language cleavage and political cleavage were further fanned by the pirate operation "Freedom Radio", which bitterly and somewhat unjustly attacked the Afrikaner and the Nationalist Party for siding with the Germans. It should be pointed out that many Afrikaners were supporters of Smuts, though many of them resented helping England a mere 30 years after the atrocities of the Anglo-Boer War. The numerous staged shutdowns of "Freedom Radio" made Afrikaners believe that it was run by key English-speaking staff of the SABC.

The Struggle for Afrikaans Within the SABC

The SABC Law No. 22 of 1936 stipulates that Afrikaans and English should receive equal priority in broadcasts. The first full-time Afrikaans programs were transmitted from Cape Town on October 1, 1937. Until October, 1937, English was the sole language of the SABC and subsequent resentment over political broadcasts probably made the SABC a prime target for a secret Afrikaner organization, the Broederbond.

The Afrikaner occupied a subservient economic and political position. One of the aims of the Afrikaner Broederbond was to infiltrate key positions in education, commerce, and industry. When the Nationalists came into power in 1948, the political future of the Afrikaner was secured, but it still took time before the Afrikaner could gain equal footing with the English in economic spheres. Broederbond-inspired economic institutions such as as Sanlam and Volkskas provided the vehicles whereby concerted economic action could be mobilized.

Nearly 80% of Nationalist MPs and virtually all cabinet minsters are thought to belong to the 12,000 strong Broederbond. Furthermore, the chairman of the board of governors of the SABC, Dr. P. J. Meyer, is a former chairman of the Broederbond. Members of the Broederbond in the SABC include the vice-chairman and top executives of the SABC. Commonality of membership of the Broederbond gives the government indirect control of the broadcasting monopoly.

Judged objectively, the SABC has thus far done its utmost to provide each cultural group with its own radio service. Certain recent incidents such as refusal to report Judge Mostert's revelation of the "information scandal," and repudiation of the top executives of the criticism of the personnel association of the SABC against press freedom being endangered by the proposed Advocate-General Bill, displayed the pro-government bias of the SABC management.

With regard to newscasts the SABC is fairly objective in that it receives its news from the South African Press Association (SAPA) and five international news agencies (Reuters, AP, UPI, Agence France Press and Deutches Press Agintur). With regard to news commentary the SABC has continually been criticized for presenting the official South African viewpoint. Most criticized is undoubtedly *Current Affairs*, a daily program of news commentary.

An aerial view of the SABC broadcast complex appears in Fig. 1-1.

Commercial Radio

On May 1st, 1950, a third service, the bilingual commercial service, Springbok Radio, was inaugurated. This service supplied advertising

Fig. 1-1. SABC broadcast complex in Auckland Park, Johannesburg. The FM tower is shown in the background.

revenue in addition to listeners' license fees which could be used for much needed expansions. Because of the close links between South African and American advertising, sponsors concentrated on popular programs such as daytime serials, soap-operas, "hit parade" and "top twenty." All in all, the programming is a blend of American and English commercially sponsored programs such as *Playhouse 90, Lux Radio Theatre, Tea with Mr. Green*, with in-between Afrikaans programs such as *So maak mens* (This is how you do it) and lots of popular music.

To meet the continuing challenge of television, Springbok Radio made a number of innovative changes during 1978 in order to provide both serious and lighthearted entertainment for the whole family. New innovative programs include *Fun with the Forces* and a discussion program *Top Level*. Over 200 plays were broadcast on Springbok Radio during 1978. These plays included inter alia: *A Doll's House*, by Henrik Ibsen; *The Letter*, by Somerset Maugham; *The Quests*, by Leonid Zorin.

Top recording artists are acknowledged with the annual *Sarie* awards. Worth noting is the Christmas fund sponsored by Springbok Radio for the benefit of charities. Over 14 years Springbok Radio has collected R3, 359, 324 in aid of charities.

Apart from the national commercial service, Springbok Radio, which broadcasts on medium and FM waves, was the first of three regional commercial services which started broadcasting in 1964. At present three regional services, Radio Highveld, Radio Good Hope, and Radio Port Natal, broadcast 132 hours each per week on FM.

Radio Five is a commercial service with the teenager and young adult as the target audience, which transmits mainly musical programs on medium wave from Johannesburg.

A radio service for blacks, in keeping with the policy of the SABC to provide each population group with its own service, was a long-felt need. Because of the limited frequencies available on AM, and the fact that full capacity was reached with the erection of the Paradys transmitters, a distribution service, whilst awaiting the completion of the FM network, was inaugurated for blacks in 1952. This service transmitted programs in Sotho, Zulu, and Xhosa to the black townships west of Johannesburg. The first FM transmission from the tower in Johannesburg occurred on Christmas 1961, a week before the FM service of Radio Bantu began on January 1, 1962. A radio RSA broadcaster is shown in Fig. 1-2.

An external radio service, Radio RSA on shortwave only, was started on May 1, 1966. The target areas of Radio RSA are Africa, Europe, United States, Canada, and the Middle East. Forty-two news bulletins and 21 news commentaries are broadcast daily in English, Afrikaans, Dutch, German, Portuguese, French, Swahili, Chicewa, Tsonga and Lozi. Twenty-three out of 26½ hours per week are devoted especially to Africa.

From a modest beginning in 1936, the SABC has developed into a giant corporation, broadcasting in 25 languages, for a total of 2280 hours per week on 20 full-time radio services and two television services.

RADIO PROGRAMMING

The commercial services broadcast by the SABC are the following:

● Springbok Radio, a total of 132 hours on medium wave, shortwave, and FM;

● The three regional commercial services, Radio Highveld, Radio Good Hope, and Radio Port Natal, each broadcasting on FM for 132 hours per week. Special features of this service are the mobile studios whereby special visits to various centers in the respective reception areas can be made. Radio Highveld also maintains a traffic helicopter to assist motorists to get to and from work;

● Radio 5, broadcasting on shortwave for 133 hours;

● Radio Bantu has seven commercial services broadcasting for a total of 1000 hours on FM. They are South Sotho, Zulu, Xhosa, North Sotho, Tswana, Venda, and Tsonga each broadcasting for 168 hours per week.

The noncommercial services are: Radio RSA, broadcasting on shortwave for 26½ hours per week; the English service and the Afrikaans service, each broadcasting 136 hours of "cultural" programs per week.

Afrikaans Programs

A very high standard of radio broadcasting is maintained in South Africa on the so-called cultural broadcast. An indication of the standard of broadcasting is derived from the fact that the French service of Radio RSA

Fig. 1-2. Radio KSA announcer.

was being voted the most popular shortwave station in the world by the "Club Ondes Courtes de Quebec." In 1976 and 1977, the German service of Radio RSA was voted the most popular shortwave station broadcasting in German by the Club ADDX for German shortwave listeners.

The Afrikaans service concentrates on cultural programs (such as dramas, documentaries and features), serious music, children's programs, sport and educational talks such as agriculture and women's topics. During 1978, the Afrikaans service of the SABC broadcasted 35 radio plays of which 16 were written by Afrikaans writers.

An analysis of weekly time allocated to different types of programs appears in Table 1-2.

English Programs

In order to compete successfully with television, radio has had to capitalize on its immediacy; therefore, actuality broadcasts play a major role also in the English service. The most significant of these actuality programs is *Radio Today* on which distinguished visitors from abroad appear. Topics which were covered include inter alia, the Camp David meeting and subsequent negotiations between Israel and Egypt, developments in Namibia and Zimbabwe, the "information" scandal, the death of the Pope, etc. Other types of programs covered by the English service include sport, business, science, culture, and human interest. Listener participation is sought with the program *Microphone-In*, in which listeners phone in their views on such diverse topics as euthanasia, seal-clubbing, and test-tube babies.

The English service also recognizes its obligation to make a contribution to formal education. Special programs of Shakespearian set work for the matriculation examination were broadcast. Apart from the Shakespearian plays, drama forms a regular feature of English programming. A total of 110 one-hour to one-and-a-half hour plays were broadcast during 1978. Twenty percent of these plays were written by South African authors.

During 1978, three programs per week, specially produced for Indian listeners, were broadcast from the Durban studios.

In Table 1-3 is an analysis of program time during a normal week. English service programs for the regional and shortwave are shown separately.

School Radio Service

Since 1936, the SABC broadcast special programs for schools. As from 1961 the SABC has appointed a full-time organizer who plans programs in collaboration with the National Council for Audiovisual Education. Filmstrips are available through national film productions to accompany the audio presentation. Publications as aids for teachers are circulated to participating schools. During 1978, 11, 407 teachers' guides were distributed. (SABC Annual Report 1978 p 109). At present the

Human Sciences Research Council is evaluating the use made of the School Radio Service.

A School Radio Service was introduced by Radio Bantu in 1964. The SABC emphasizes that the School Radio Service should be seen as an aid in formal and formative education. The result is that the initial programs were not exclusively subject-directed, but covered many aspects of interest. This apparently confused the listeners. An investigation was consequently launched in 1966, with resultant changes in 1967 in that separate senior primary standards received specific lessons which were intended only for that standard.

From 1973 onward, the School Radio Service offered programs for senior primary, junior primary, and secondary pupils, as well as for students at the training colleges. The Zulu service of Radio Bantu

Table 1-2. Analysis of Weekly Time Allocated to Various Program Categories.

Programs	Hours	Percentage
National Broadcasts		
Light music		
• Record programs 15 hours, 11.1%)		
• SABC productions 1¾ hours, 1.3%)	16¾	
12.4		
Classic and light classic		
• Record programs 10½ hours, 7.7%)		
• SABC productions 5¼ hours, 3.8%)	15¾	11.5
Request programs	12	8.8
Actuality and journal	18	13.2
News and news commentary	10½	7.7
Sport	9	6.6
Religious programs	8¾	6.4
Youth and Children's programs	6¼	4.6
Drama	5¾	4.2
Entertainment, quiz and Competition programs	5	3.7
Agricultural programs	2½	1.8
Discussions and features	2¼	1.7
Weather reports	2¼	1.7
Stock exchange, trade bulletins and economic news	2	1.5
Literary and cultural programs	1¾	1.3
Program publicity	1¼	0.9
Parliamentary reports	1	0.7
Talks	¾	0.6
	121½	89.3
Regional and shortwave broadcasts		
Magazine, request and sports programs	9	6.6
Market reports and meat prices	2¼	1.7
Religious services	1¾	1.3
News	1½	1.1
Total	14½	10.7
Grand total	136	100.0%

Table 1-3. Analysis of English Program Time During a Typical Week.

Type of Program	Time Hours. Minutes)	Percentage
National broadcast		
News and new commentaries	10'15"	7.74
Radio Today	9'00"	6.79
Talks and magazines	7'09"	5.39
Actuality	8'40"	6.55
Agricultural broadcasts	2'30"	1.88
Sport	9'00"	6.79
Literary programs	2'00"	1.51
Drama and features	12'00"	9.07
Youth and children's programs	2'15"	1.71
Variety and quizzes	1'48"	1.36
Service programs	3'10"	2.39
Religious broadcasts	6'55"	5.23
Indian programs	1'40"	1.27
Music (light)	22'14"	16.79
Music (serious)	21'59"	16.47
	120'35"	90.94%
Regional and shortwave broadcasts		
News	8'05"	6.16
Talks and magazines	30"	0.37
Agricultural broadcasts	1'30"	1.14
Sport	55"	0.68
Indian programs	15"	0.19
Music	40"	0.51
	11'55"	9.06%
Grand total	132'30"	100.00%

introduced an over-the-air university in which subjects such as psychology, sociology, geography, etc., are discussed.

The aim of Radio Bantu, apart from the School Radio Service, is also entertainment and education as well as influencing. A good example of such programs is "Impilo Yethu" (Our Health), which originated during 1974 on the Xhosa service of Radio Bantu. Songs and melodies are used to assist in remembering facts in the programs. The way in which the programs have been compiled takes due consideration of the vocabulary and cultural forms of the Xhosa. Research conducted by the Human Sciences Research Council indicate that 80% of the population of Transkei had heard the program on the radio, 51% of the respondents remembered the contents and 18% had carried out specific actions with regard to the program (Van Vuuren D, Puth G. and Roos D. The Scope and Impact of the health guidance program *Impilo Yethu* of the Xhosa Service of the SABC, - Comm. N-3, 1977).

The Bantu Radio Service of the SABC has as its aim education as well as influencing. Radio has tremendous informational value which is used to make the various ethnic groups aware of their identity and to appreciate that which is beautiful in their own culture. Such a goal would have been laudable in any country but South Africa. Providing each cultural group with its own radio service is seen as yet another despicable result of the policy of separate development in that it perpetuates distinction and reduces commonality.

One of the functions of the mass media is cultural transmission. The SABC would indeed be foolish if they did not use the potential power of radio to influence, especially if such attempts at influencing leads to a reduction of conflict and better understanding of its various peoples.

Music

The SABC Symphony Orchestra gave its first public performance in 1954. Since then the National Symphony Orchestra has performed regularly in Johannesburg and Pretoria and elsewhere. Lunch-hour concerts by the symphony orchestra became very popular in Johannesburg. During 1978 a total of 87 public performances were held (SABC Annual Report, p 95).

Since music comprises 50% of all radio time, the SABC feels obligated to stimulate not only serious but also light music. The Department of light music of the SABC has an important function with regard to the production of South African light music. The annual SABC music competition draws top young performers and is regarded as the prestige musical event of the year.

The SABC has formed a junior symphony orchestra which offers an opportunity to young orchestral musicians to acquire experience in orchestral playing.

RESEARCH

The SABC has its own research department whilst other research is contracted out to commercial companies. As from 1970, the Institute of Communication Research of the Human Sciences Research Council conducts research for the SABC on the effects and patterns of television viewing. Listener and viewer research is conducted by the SABC throughout the year amongst all population groups.

On average more than 9.9 million adult listeners tune in daily, Monday to Friday, to the various radio services. The combined daily listenership to the various services in Bantu Languages was an average of 4,891,000 adults for 1978 (SABC Annual Report, 1978, p 125).

A television viewers' panel, representative of all population groups (excluding at present the blacks), was established in order to gauge the attitudes and preferences of television viewers. The findings obtained from the panel of television veiwers are made available to the relevant program departments.

TELEVISION SERVICES

The Broadcasting Amendment Act of 1949 gave the SABC the ultimate authority over television in South Africa. However, because of technical, financial, and other reasons, it took until 1976 before the first regular television programs were broadcast.

The Television Controversy

It is true that then Minister of Posts and Telegraphs Albert Hertzog strongly opposed the introduction of television on the grounds of the corruptive influence of the medium. However, the real reasons for nonintroduction are to be found in the following:

● A network of FM towers which could be utilized in television transmissions had to be erected. This paved the way for the introduction of television since the SABC could not contemplate television before radio could be supplied to the whole population.

● A suitable economic infrastructure, and expecially a well-developed electronic industry, had to be established before introduction of the medium.

●South Africa is large and sparsely populated, with geographical barriers. The SABC had to satisfy itself that television was technically developed to such an extent that these barriers could be overcome.

●Programming and production costs are prohibitively high. A sufficiently large portion of the population had to be able to afford television sets as well as the relatively high television licensing fee.

● The SABC had to be sure that it would be able to keep up programming to satisfy the insatiable need of the medium. A genuine concern was cultural and sociological rather than political in nature. Cultural invasion due to importation of programs and disruption of family life were concerns expressed. Control of the government over the SABC nullifies the argument that the government really was afraid of the "dangerously liberalistic ideas" to be set loose upon the population through British and American "canned" programs. The same control exercised over screening of foreign movies could, in fact, apply to television programs.

After a long period of preparation and training of television technicians, mainly by staff recruited from the BBC, the first trial television transmissions started in October 1975. The first single-channel service in Afrikaans and English started on January 5th, 1976. A total of 37 hours per week, divided equally between Afrikaans and English, is transmitted.

The SABC Television Center in Johannesburg is a massive 8-story building on six hectares of land. The building includes two studios of 250 m, four studios of 450 m and one of 900 m. Thirty-four transmitters together with an extensive relay system will eventually bring television within reach of everybody in South Africa. Five mobile units, one in each major city, are used to do on-the-spot coverage. A view of a television control room appears in Fig. 1-3.

In 1977, South Africa had the largest number of television sets in Africa, 670,000 (S. A. Yearbook, p. 821). This number is expected to increase dramatically with the prospective television service for black viewers on Channel II.

Television Programming

Approximately 60% of programs broadcast on SABC-TV are manufactured locally. For the remaining 40% SABC had to rely on overseas purchases. The PAL color system and technical quality of SABC television is undoubtedly high. Unfortunately, the same cannot be said of local program quality, probably due to lack of experience with the medium and adequate training facilities in schools of communication.

According to the SABC annual report (p. 59), program types comprise the following percentages:

Drama	24%
Documentaries	15%
Sport	14%
Variety	12%
News	11%
Magazines	9%
Children	7%
Serious music	4%
Religious	4%
	100%

The most popular programs with South Africans are the news, sport programs, and actuality (Fig. 1-4). Despite cooperation with overseas networks the SABC faced new challenges in terms of sources of information for television news. A corps of specialist reporters, properly acquainted with their subjects, had to be developed. Part-time correspondents have been appointed in key African states and agreements reached with large news organizations. The SABC has 21 news offices throughout South Africa and one in London. Additional coverage is supplied by 1200 part-time local and overseas correspondents.

Fig. 1-3. View of a television control room.

Fig. 1-4. Fully equipped vehicles for outside broadcasts.

The SABC Television News received 47 satellite feeds from overseas during 1978. The possibility of receiving such feeds on a daily basis was discarded because of the high cost involved. SABC in Johannesburg also functions as a center for overseas networks. During 1978, facilities were made available to BBC, ITN, Visnews, UPITN, and the American networks NBC and ABC for the transmission of 219 satellite feeds (SABC Annual Report 1978 p. 63).

A one-channel television service catering to Afrikaans and English viewers has led undoubtedly to a greater understanding between the two language groups. Frequent programs about other ethnic groups and appearance of coloreds and blacks on television has created a noticeable understanding of other races. Blacks and coloreds frequently conduct religious services on television. Since South Africans are religious, such appearances foster respect for other races.

Chapter 2
Broadcasting In Poland

Prof. Stanislaw Miszczak
Warsaw University

In June, 1924, the Ministry of Industry and Trade invited tenders for broadcasting services in Poland. The ministry gave August 31, 1924, as the dateline for the applications for radio concessions from interested persons and legal bodies.

A dozen or so firms and corporate bodies from Poland and abroad sent in their tenders. Among them were the Polish Radiotechnological Society (PTR), the largest firm of that kind in Poland in those days, and an unknown company, "Polish Radio" (Polskie Radio), Ltd.

BIRTH OF POLISH BROADCASTING

The Polish Radio Company came into being on February 5, 1924, with the purpose of establishing Poland's first radio broadcasting house to be financed entirely from national sources. Eventually, on August 18, 1925, the Polish Radio Company won the tender and obtained the concession for outside broadcasting throughout the territory of the Polish Republic.

However, Poland's Radio broadcasting was already an accomplished fact before, owing to the Polish Radiotechnological Society. On February 1, 1925, the society put into operation the country's first broadcasting center in Warsaw. The center broadcast its opening program at 18:00 hours. This date is frequently taken as the birthday of Polish broadcasting.

Shortly before it officially went into service, the center was registered by the Ministry of Industry and Trade as an experimental factory site, which didn't clash with the regulations then in force.

The center broadcast daily both music and verbal programs between 18:00 and 19:00 hours, and the most prominent people were invited to take part in them. On February 10, 1925, the center began regular broadcasts of the weather report of the State Meteorological Institute and the news bulletin supplied by the Polish Telegraphic Agency.

The Polish Radiotechnological Society's decision to set up a radio center was backed up by certain tactical reasons. One can hardly deny that the society, by giving its center a go-ahead, intended to draw the government to its side.

As the competition for who would run the broadcasting services in Poland was dragging out inconclusively, the Polish Radiotechnological Society made known that it was running out of funds for its broadcasting venture and that it would, therefore, discontinue the broadcasts.

In such circumstances, various welfare organizations affiliated with the central committee of the Polish Radiotechnological Associations went into action. On June 10, 1925, the committee appealed to all amateur radio operators and radio sympathizers to volunteer financial contributions to ensure continuation of the broadcasts by the Polish Radiotechnological Society Center. The appeal released a tremendous amount of social initiative and proved very effective. The broadcasts were continued.

The programs sponsored by the central committee of the Polish Radiotechnological Associations were broadcast three times a week on top of a weekly program sponsored by the Polish Radiotechnological Society and a fortnightly one sponsored by "The Republic" (Rzeczpospolita) daily.

Upon receiving the concession, the Polish Radio Company, Ltd., had no broadcasting center of its own nor any broadcasting experience. Therefore, it chose to cooperate closely with the central committee of the Radiotechnological Associations, which by the time had considerably expanded its activities in the field of organizing radio broadcasts.

As a result of negotiations, the Polish Radiotechnological Society signed a contract placing its center at the disposal of the Polish Radio Company, while the latter transferred the right of exploitation of the center to the central committee of the Polish Radiotechnological Associations. Moreover, the Polish Radio Company offered the committee a regular, even though insignificant, financial subsidy.

Under the contract, the Polish Radiotechnological Society Center, which broke off its services on August 18, 1925, reactivated the broadcasting, which continued uninterruptedly between November 26, 1925, and March 14, 1926, when the Polish Radio Company began testing its own broadcasting center. The central committee of the Polish Radiotechnological Associations, being a welfare organization, was then the actual organizer and sponsor of the broadcasts, an unprecedented and noteworthy fact in radio history in Europe. The programs broadcasted at that time totalled 103 hours and featured 76 soloists.

On November 26, 1925, when the center, managed by the central committee of the Polish Radiotechnological Associations re-inaugurated its broadcasting, a group of prominent artists were invited to perform. From then on, the center's broadcasting was getting more and more professional, winning respect of both listeners and artists in Warsaw. Soon regular programs for children went on the air. At that time also, on November 29, 1926, the first radio play, "Warzawianka," by Stanislaw Wyspianski, was broadcast.

ORGANIZATION OF BROADCASTING IN POLAND

Under the concession of August 18, 1925, the Polish Radio Company, Ltd. was obligated to build and put into operation a broadcasting center in

Warsaw (within six months of the concession issuance date) and in Krakow (within 12 months). When the number of licensed owners of radio sets in Poland topped 60,000, the company's obligation was to build a third broadcasting center on a site to be determined by the general management of Post and Telegraph Services. As the number of radio owners expanded, successive centers were to follow.

The government also committed the broadcasting concession holders to form a joint stock company with assets amounting to 1,250,000 zlotys. It was formed in May, 1926, and its founding assets were divided into 100 zloty shares. The statute of Polish Radio as a joint stock company was announced in the Government Official Gazette of 1926. The Government reserved to itself 40% of the company's shares, leaving the remaining 60% in private hands.

The company received the concession for the construction and operation of radiotechnological installations for the period of 10 years. The concession holders were authorized to collect fixed license fees from owners of radio sets.

On July 30, 1929, the Polish Radio Company received a renewed concession from the Ministry of Post and Telegraph Services for a period of 20 years, that is, up to 1949.

The basic obligations of the concession holders included the mainte- nance and sustained operation of the existing broadcasting centers, the construction of the new ones, independent organization and financing of the programs, introduction of new technologies and improvement of those already in operation. These obligations were all designed to ensure that the entire territory of the state is covered by the range of broadcasting centers powerful enough to be received by detector sets. The broadcast- ing installations were to expand so as to ensure services to the regions that would prove the most profitable in the view of the density of their population and the cultural standard.

The Polish Radio Company received 80% of the revenues from the licenses paid by the listeners through the agency of post offices. Post offices also carried out the operations connected with the issuance of subscriber licenses and installation of receiving sets.

Subsequently, the distribution of the revenues from the subscriber licenses underwent a relevant change. In the period between October 1, 1929, and April 30, 1932, the state treasury received only 15% of the revenues, instead of the prior 20%. In the period from October 1, 1932, to September 30, 1933, this share amounted to 17.5%, while beginning October 1, 1933, up to the expiry of the concession, it was to run at 20%.

The Polish Radio Company, Ltd., was headed by a director in chief. The company's authorities were: the general assembly of shareholders, the managing council, with one third to the members appointed by the minister of post and Telegraph Services, and the auditing commission. The institution, in its daily work, was run by three directors heading three sections: programs, technological, and administrative. The sections, in turn, broke down into departments.

In order to make the company's work more efficient, the Central Program Council was established in 1931. It was made up of nine members and a chairman. The chairman and five of the members were appointed by the minister of Post and Telegraph Services. The remaining members were approved of by the minister of Post and Telegraph Services on the motion of the director in chief. The main task of the program council was the program control, approval of general schedules, and analysis of all programs connected with the production of programs broadcast by Polish Radio.

In 1935, the government bought out a majority of private-held shares of the Polish Radio Company in the drive to enlarge its influence on the Polish broadcasting services. As a result of this move, the state treasury (working through the ministry of Post and Telegraph Services) boosted its company stock from 40% to 95.8%. Thus, Polish Radio was practically nationalized, though the company did not change its legal status of a private enterprise as it continued to bear all charges and taxes, like other private firms. At the beginning of 1938, the state treasury share in the polish Radio Company stock increased from 95.8% to 97.1% and amounted to 2,913,000 zlotys.

RADIO PROGRAM DEVELOPMENT

During the years between the world wars, Polish Radio broadcast one nationwide service. Its basic components were the programs produced by the central broadcasting house in Warsaw. The service was broadcast through the long-wave transmitter in Raszyn and rebroadcast (by wire or wireless) by medium-wave transmitters in other broadcasting regions. The regional broadcasting centers also used their own transmitters for scattered local programs. One exception was Warsaw, which, beginning in 1937, had its own local service.

The basic components of the programs in the period between the wars was music, mainly played from gramophone (phonograph) records, and, secondly, literary programs. Programs of the informative character were initially made up only of news bulletins and various announcements. Later, topical reportage was gradually expanded.

It is worth noting, at this point, that Polish Radio, in the years 1930-1936, broadcast a big number of advertising interludes, which accounted for at least 3.1% of the total program time. The revenues from advertisements amounted to 9% of Polish Radio's total income, which was quite a substantial share. However, the advertising programs displeased the organization of publishers of journals and newspapers who believed that such programs decreased the incomes of the press. This resulted in the little interest shown by that organization in popularizing broadcasting among broad masses of readers. The obstacles to cooperation were solved by a bilateral contract signed on March 31, 1937. As of that day, Polish Radio discontinued broadcasting advertisements.

The share of various program components was different in the nationwide and the local broadcasts. The latter contained more music. The situation was the opposite in the case of verbal programs.

As the result of the growing number of transmission hours and increasingly complex programs, the number of program performers was steadily growing. In 1926, 640 performers took part in artistic programs (music and literary), while in 1938 their number soared to 4,300.

In 1935, Polish Radio founded its own symphony orchestra in Warsaw, made up of 52 musicians. The orchestra gave its first concert on the air on October 2, 1935.

The foundation of an independent symphony orchestra was an event of great importance for Polish broadcasting. The orchestra issue had been debated for many years. For a long time, Polish Radio had used the services of an orchestra composed of musicians from the Warsaw Philharmonic Society. However, the duality of the musicians' activities failed to guarantee proper results in repertoire preparations either for Polish Radio or for the Warsaw Philharmonic.

In 1938, Polish Radio founded four new orchestras in its regional broadcasting centers. Moreover, various popular music bands were contracted to perform on the air.

The most significant place in light programs was occupied by the Small Orchestra of Polish Radio, established in April, 1935, with the aim of popularizing light music. The orchestra, initially composed of 12 members, each of whom played several instruments, very shortly became immensely popular with the radio listeners.

Its popularity grew even more when Polish Radio organized, in the Bristol Hotel in Warsaw at the beginning of 1936, regular Sunday *Five O'Clock at the Mike*, a popular variety event during which the mentioned above Small Orchestra, renowned singers, instrumentalists, humorists and narrators were performing. In the period between May 1, 1935, and May 1, 1938, the Polish Radio Small Orchestra performed 1,000 times on the air.

Apart from the national program, Polish Radio, from March 1, 1936, on, also broadcast the programs abroad, mainly for the Poles living overseas. These broadcasts were beamed by means of a 10 kW SPW transmitter on the wavelength of 22 meters. The transmitter was leased from the Ministry of Post and Telegraph Services under a special agreement.

Initially, the shortwave broadcasts took place three times a week between 17:30 and 18:20 and later between 18:30 and 19:30 hours. Beginning on May 3, 1936, two-hour broadcasts were introduced on Sundays and other holidays between 17:30 and 19:30 hours. The broadcasts consisted of news bulletins and talk about Poland, its past, historical relics, social life, etc. The center also broadcast a separate news bulletin in English, with an eye on the Polish Americans.

On October 3, 1937, the second shortwave transmitter entered into operation. Its power output was 2 kW, the wavelength was 26.01 meters and the calling signal SPD. Then the transmission time was extended to three hours a day on Saturdays, Sundays and other holidays. At that time the broadcasting hours were changed so as to reach American listeners at 18:00. The broadcasts for Polish emigrants in Europe were also introduced, using Polish Radio long- and medium-wave transmitters.

POLISH RADIO BROADCASTING DURING WORLD WAR II

Immediately after the German invasion of Poland, the invaders issued an instruction ordering the civilian propulation to turn over the radio equipment in private possession, notably radio receivers, to the authorities in occupation, under the penalty of death.

However, the penalty of death failed to discourage Poles from seeking contact with the world. There were thousands of people in Poland who, notwithstanding the German threats, didn't abide by the order and kept their radio receivers. Moreover, the threat of illegality didn't refrain many others from constructing radio sets on their own. The components for such makeshift radios were clandestinely carried away by Polish workers from German-controlled radio factories.

The expanding resistance movement in the occupied country and the political organizations springing up in the underground were more and more in need of a communication means to be able to coordinate their operations against the Germans. They received the indispensable radio equipment both from parachute drops and from their own underground radio workshops.

The handling of the transmitters was a very dangerous job calling for great responsibility, because the Germans employed goniometry on a large scale to track down underground radio communications stations. Thus, the stations operated by the resistance movement had to be very mobile. A different matter was the protection of larger underground transmitters that were used by large resistance units or were destined for use in the event of mass-scale operations against the invader.

To make it difficult for the Germans to detect the transmitters operated by the Polish resistance movement, the singular "compensation" method was employed. Owing to the application of the compensation device in the transmitter antenna system, it was possible to achieve a total suppression of ground signals along the antenna plane, which made the station impossible to locate by means of goniometric equipment.

At the time when Hitler attempted to isolate Poland within a wall of silence, Polish speech began flowing on the air. Various radio stations of the states, members of the anti-Nazi alliance, broadcast programs in Polish. These broadcasts were the main source of information of the whole underground press in Poland. The broadcasting stations performed the momentous task of mobilizing people for the struggle against the enemy, and of pointing up to them the prospects of victory and the vistas of lasting peace.

The first to broadcast in Polish was the BBC, London. It was followed by the French station a few months later. During the four years of the war, the BBC Polish services expanded immensely. At the same time, the radio station "The Dawn" (Swit.) which the British put at the disposal of general Wladyslaw Sikorski, began operation. The programs of that station were chiefly composed of information that the leaders of the underground struggle in Poland transmitted to London by various channels. In August, 1942, the BBC began an American service in Polish, too.

One of the most important events in Polish broadcasting was the report on Rommel's retreat from El Alamein. The news reached listeners in Poland a few minutes after the special announcement was released in Cairo and before it reached listeners in England. Shortly afterward, the Poles learned in the same way about the recapture of Tobruk by the 8th Army, which included also the Polish Carpathian Brigade.

At the end of 1944, the Lublin broadcasting station transmitted up to six hours a day on the wavelength of 336 meters, then switching to 224 meters. Its services were composed mainly of information and announcements, as such programs were most in demand at that time. Besides, artistic and music programs were hardly feasible, because of the shortage of technical facilities.

On December 31, 1944, foreign-language services were initiated. As a result, the world could learn about the session of the National People's Council (KRN) at which the provisional government was established. The foreign-language services were broadcast by means of a 7.5 kW transmitter.

The foreign-language services were of tremendous political importance at that time, as they were the main—and frequently the only—source of information for the world on what was going on in Poland.

From the moment those services began operation, there was not a single day that the Polish broadcasts from Lublin would not be quoted by radio stations around the world. A number of newspapers and radio stations, such as Reuters or the BBC, recorded the Polish broadcasts on discs for subsequent popularization in the papers or reproduced them by their own broadcasting means.

On February 1, 1944, the BBC radio stations also broadcast the news of the assassination of the SS and police commandant in the so-called Warsaw district, General Franz Kutshera. The news, also broadcast by the Dawn Radio station, reached listeners almost immediately after the operation, thus making the Germans aware of the fact that the Polish underground's communication with London was organized perfectly.

One particularly vital broadcast, both because of its military importance and because of the circumstances, told about the landing of Allied Forces in Normandy. The news was broadcast to Poland 15 minutes before the actual landing operation took place; thus, it was the world's first news about the landing that was broadcast by any world station, including the BBC.

On June 25, 1941, Radio Moscow began broadcasting regular services in Polish. The initially modest Polish section of that station soon grew to a dozen or so people and closely cooperated with outstanding Polish Communists. In October, 1941, the Polish section was evacuated to Kuvbishev along with many other Soviet radio agencies. From Kuvbishev, informative, verbal and music programs were broadcast, addressed to the Poles scattered all over the Soviet territory, as well as for those in the occupied country.

RADIO SERVICES AFTER THE WAR

Evaluating the road which Polish radio broadcasting has traversed since the conclusion of World War II, one can hardly do without a brief understanding of the condition of radio services in the years 1944-1945. It is impossible to understand the problems that the Polish broadcasting has faced over the past 30 years without realizing what happened during the occupation. No other branch had been so thoroughly damaged by the invaders as the radio services and facilities were. Poland had virtually no surviving radio facility that could resume its duties immediately after the liberation of the country. An overwhelming majority of radio facilities had been either thoroughly plundered or, usually, blown up.

Likewise, the Polish Radiotechnological industry had been ravaged by nearly 98%. Some factories had been destroyed as a result of military operations, others had been looted by the retreating invaders.

This is roughly a balance sheet of the material losses sustained by Polish broadcasting, which had to be faced in 1945. But there were also other losses. The criminal Nazi occupation had drained Poland of qualified personnel. The task of the reconstruction had to be coped with from the beginning and with literally bare hands. In those circumstances, any piece of radio equipment salvaged from the ruins was important; any specialist, any means of transport and communication were precious. Reborn Polish broadcasting had to be endowed with a proper organizational structure attuned to the changing requirements, program and technological capacities.

When one looks back to those years from the point of view of the present times, one can hardly believe that, given the hopeless technological situation and complete lack of materials, Polish specialists nevertheless did bring back to life all the radio facilities that had been thoroughly destroyed during the war. It took a great ardor and persistence to undertake this job and carry it to a successful completion. All those people were propelled by the awareness of the fact that their great objective, the reconstruction of the barbarously destroyed Polish Radio broadcasting, contributed to the foundation of the new Poland. At that time, the entry into operation of every new broadcasting center was great national event.

ORGANIZATION OF RADIO AND TELEVISION SERVICES IN POSTWAR POLAND

The first postwar legal act concerning radio broadcasting was a decree issued by the Polish Committee for National Liberation on November 22, 1944. Under the decree, the state enterprise "Polish Radio" was established. Its tasks included the construction and operation of radio transmitting installations and radio relay centers, as well as the producing of radio programs for broadcasting. The enterprise enjoyed a legal status and its property was separate from the state treasury property.

The enterprise was headed by a director, appointed by the Polish Committee for National Liberation on the motion of the head of the Propaganda and Information Department. The technical supervision rested in the hands of the director in chief of the Communication, Post and Telegraph Services Department. Organizational matters, as well as those concerning repertory, administration and finances, were controlled by the head of the Propaganda and Information Department.

With the closing down of the office of the Minister of Information and Propaganda, the supervision of Polish Radio was taken over by the prime minister. Only the supervision of the cultural and artistic repertory was in the hands of the Minister of Culture and Arts.

In 1948, Polish Radio had four departments: administrative-commercial, technical, finanicial, and programs. There were also nine regional boards of directors.

Under the law of February 4, 1949, the Central Broadcasting Office (CBO) was established as the superior office for radio and television broadcasting for mass reception. The establishing of CBO marked the end of the first postwar stage of the reconstruction of Polish braodcasting and the beginning of a new stage of expansion.

The CBO tasks included:

● plans   for the development of radio services, television and broadcasting centers

● program,   organization, technical, administrative and financial control over enterprises operating radio and television broadcasting equipment (wireless and wire)

● the issue of licenses for the establishment and running of radio and television transceiving centers for mass use and of licenses for possession and use of radio receiving equipment (wireless and wire)

● representing Polish radiophony and television on the international arena

The CBO was subordinate to the prime minister. It was headed by a president who was appointed by the President of the People's Republic of Poland, on the motion from the prime minister.

The CBO was composed of: the secretary general Office, planning and coordination office, technological progress office, personnel office, finance office, control office and an independent department for program studies and supervision.

Subordinate to CBO were the state enterprise Polish Radio and the state enterprise for the Development of Radio Services, established by the prime minister on December 31, 1949, which took over from Polish Radio all the matters connected with wire radiophony in general and the setting up of local and megaphone equipment, as well as the running of repair workshops. The two latter tasks were assigned to two boards of directors, one for wire radiophony, expanding radio services by installing and maintaining wire broadcasting centers, and the other for radio receiving equipment, which was responsible for installing local radio equipment and megaphones (public address system) for the purpose of mass events. That board of directors also supervised radio receiver service stations.

In order to improve the quality of radio programs, the prime minister, on May 17, 1950, appointed the program council to assist the president of the CBO and decided on its members and regulations.

The subsequent legal act of fundamental value to the organization of radio services was the decree of August 2, 1951. The decree established the Committee for Radiophony "Polish Radio" as a government organ concerned with producing and broadcasting radio and television programs for mass reception. Simultaneously, the CBO and the state enterprise "Polish Radio" were closed up.

The following former CBO tasks passed to the Minister of Post and Telegraph Services (now the Minister of Communication):

● construction and operation of radio broadcasting centers for mass reception

● plans for the development of radio services

● wire broadcasting for mass reception

● installation and use of radio and television (wire and wireless) receiving equipment for mass reception

The remaining tasks of the CBO passed to the Committee for Radiophony "Polish Radio."

The development of television broadcasting resulted in the transformation of the Committee for Radiophony "Polish Radio" into the Committee for Radio and Television "Polish Radio and Television," under the law of December 2, 1960. Under this law, the committee is the central organ of the state administration for radio and television broadcasting for mass reception. The act is still valid, at present.

The committee's sphere of activities includes, in particular:

● marking out, on the basis of national economic plans and resolutions of the Council of Ministers, of the general lines for the development of radio and television services for mass reception, both in respect of the programs and organization, and consultation with the Ministry of Communication as regards the technical aspects

● exclusive production and broadcasting for mass reception in Poland and abroad of radio and television informative, political, music, literary, educational and other programs

● investments in the field of construction and exclusive operation of radio and television broadcasting houses for mass reception

● advancement of artistic, literary and scientific production, as well as of educational activities in the field of radio and television services, in consultation with the ministries concerned

● management of international cooperation in the field of radio and television services and representing the Polish radio and television on the international arena

● working out and presenting to the respective state organs the conclusions and proposals concerning production of receiving sets and radio and television broadcasting equipment, organization of the sale and repairs of receiving sets, and the license payments for using the receiving equipment.

The committee is composed of the president and his deputies (appointed by the prime minister) and members appointed from among the representatives of the central state organs and from among the responsible committee employees.

The statute, granted to the committee by decree No. 42/61 of January 24, 1961, of the Council of Ministers, established the committee's operating principles and organization. The new statute was conferred upon the committee under the decree No. 227/69 of December 31, 1969, issued by the Council of Ministers. The new law retained in force the heretofore range of the committee's activities, as defined in the earlier law of December 2, 1960, introducing some changes only in connection with the development of radio and television in the years 1961-1969 and with the plans for the development in the years 1971-1975.

Under the statute, the managing and executive organs of the committee are: the president, deputy presidents, director-general, and program directors. The presidium is entitled to act on the behalf of the Committee in Between-the-Committee sessions, and it resolves current matters.

The president of the committee administers the work of both the committee and the presidium, runs the committee's executive machine, represents the committee outside, reports on the work to the government. Over 8000 people were employed in Polish radio and television in 1977.

RADIO PROGRAM DEVELOPMENT

Programs developed along with the progressing constructions of radio projects. The task was as difficult as the erection of radio facilities, which was the job of technicians.

The program people were facing the task of uniting the entire society in the postwar reconstruction and of popularizing the idea of a new social system in Poland. Owing to its suggestiveness and straightforwardness, the radio program played a tremendous role in this respect.

In the first postwar years, Polish Radio broadcast one nationwide (central) program on the longwaves, beginning in 1949. Its initial broadcasting time was between five and eight hours daily. At the end of 1945, the central broadcasting time was expanded to 16 hours during each 24-hour period. In 1945, the central broadcasting network transmitted 2747 hours of programs, and in 1946, 4702 hours.

On October 3, 1949, Polish Radio put into service a parallel central broadcasting network on medium waves, which was a big organizational and technological achievement. With the start of Radio 2, Polish Radio broadcasting time expanded further to a considerable extent. In 1949, the total broadcasting time of Radio 1 (long waves) and Radio 2 (medium waves) amounted to 8834 hours, and in 1951, 10,250 hours—nearly 400% of what it was in 1945.

On March 1, 1958, Polish Radio inaugurated Radio 3 or VHF. The programs of this network were broadcast from Warsaw and rebroadcast from Katowice and Opole. The further development of the Radio 3 central broadcasting network took place after 1960. However, it was not until April 1, 1962, that Radio 3 became fully independent as a nationwide transmission network.

The Polish Radio broadcasts are largely produced by Warsaw braodcasting centers. In 1949, they produced 6900 hours of programs for central broadcasting, while the corresponding figure in 1967 was 14,016 hours. The remainder of programs broadcast centrally are produced by regional broadcasting centers. In 1949, they produced 1834 hours of programs for central broadcasting, and in 1967, 3137 hours, 22.0% and 18.2%, respectively.

The participation of regional broadcasting houses in the production of nationwide programs is not only a matter of prestige, but also of the set up of creative circles in capitals of provinces, a process now going under the fashionable name of the "decentralization of cultural creation."

In the period of time under review here the individual central broadcasts differed not only in the wavelength, but also in the character of their contents. Radio 1, having, as it does today, the largest reception range, broadcast mainly political-informative, popular-entertainment and educational broadcasts, as well as a majority of programs for children. Radio 2 also carried major political-informative broadcasts (reportage, commentary, correspondence), and educational programs for adults with second level education.

The artistic and light programs of Radio 2 contained much more contemporary items compared with those on Radio 1 (radio plays and music broadcasts, including jazz). On the other hand, Radio 3 broadcast mainly music and light programs. Initially, these programs were rather exclusive, as they were aimed at the listeners with an above-average musical and literary education. These broadcasts contained many vanguard and difficult items, such as modern songs, punctuational and other sorts of vanguard music, as well as modern radio plays, mainly light ones.

In 1965, Radio 3 broadcasts were simplified both in the formal sense and with regard to contents. In effect, the more difficult programs on serious music were nearly completely eliminated, as were those concerning the techniques of modern composition and other musicological programs on classical music. Classic and modern jazz music broadcasts were left intact. Also the cycle of operatic music programs and many more of light music were introduced. The light music broadcasts ranged from pop songs to rock. Also, many literary programs were introduced, including many novel series, some of them with a dozen or so installments.

Programs broadcast by Polish Radio regional centers have been and still are a substantial component of the national program. The development of such regional programs in Poland is very dynamic. In 1947, all local broadcasting centers produced a total of 2428 hours of programs, whereas 10 years later, in 1967, their program time increased to 14,085 hours, an almost sixfold rise. In 1967, local programs accounted for over 80% of the total nationwide emission time, an unprecedented percentage in the history of radio services in Europe. However, it should be noted that the question of the number and dimensions of local programs is one of the most controversial problems among the radio issues.

The structure of local broadcasts is different from that of nationwide broadcasts, as the local broadcasts are adjusted to the social, economic and demographic conditions of a given region. The backbone of these broadcasts are political-informative and economic programs.

Alongside national broadcasts, Polish Radio has been running services for foreign listeners in several languages almost since the liberation of the country. These services are received by the listeners in

France, Belgium, and Switzerland, West and East Germany, Italy, Spain, Great Britain and Scandinavian countries. There are also broadcasts for countries in Africa, Latin America and other parts of the world. Polish Radio foreign services also include broadcasts in Polish for Poles living abroad and those staying abroad temporarily, among them Polish seamen, opensea fishermen and members of polar expeditions.

As estimated number of the people of Polish ancestry living abroad is 15 million, including six million Polish Americans. For this group of listeners, Polish Radio broadcasted an average of 13 hours a day in 1968. Listeners abroad sent in nearly 30,000 letters in that year. At present, the number of letters from foreign listeners exceeds 60,000 a year. Statistical surveys, carried out by various Western institutions and associations, indicate a large popularity of Polish radio services for abroad.

Informative broadcasts were the basic program components during the period after the liberation. In 1945, these broadcasts accounted for 41.0% of the overall nationwide broadcasting time.

A particular item among the informative broadcasts was *The Search for Families*, emitted several times a day. At that time, Polish Radio was practically the only Polish institution to assume responsibility for this activity. Owing to these broadcasts, thousands of people reunited, preserving lifelong gratitude for the institution which had helped them in a difficult situation.

But this was not the only function which the Polish broadcasts had to fulfill at that time. An issue of essential importance was to develop various forms of education, that would reach a broad spectrum of listeners. The demand for such services was tremendous. The war had torn millions of Poles away from schools. The social reconstruction of the countryside and the industrialization of the cities called for expert information on the newest technological achievements. Didactic assistance was also needed in schools that were mostly stripped of textbooks and had a shortage of teachers.

The education through the radio began with the cycle entitled *Listen and Learn*. Next in the years 1946-1947 the Radio People's University was established. Finally, on September 1, 1948, the Radio University was established, shortly becoming the largest school in the country. The university was sponsored by Polish Radio and the Ministry of Educations.

Educational broadcasts also included informative talks of a practical character. They concerned professional guidance, work competition, work hygiene and safety and the struggle against occupational diseases.

Broadcasts for schools also figured highly. One of the great merits of Polish Radio was the introduction of kindergarten broadcasts, one of the first ventures of the kind in the world. These broadcasts were meant as methodological aid for kindergarten tutors and provided materials for the lessons. Musical education played a particularly important part; rhythmic games, songs, folk dances, playing various instruments, etc.

In the first postwar years, Polish Radio laid heavy stress on the so-nalled mass broadcasts. Radio teams visited dayrooms, factory halls and other places, giving music and literary performances there and popularizing current social and political events. These ventures gave rise to the broadcasts entitled *Saturday After Workday.*

Another great accomplishment of Polish Radio were music broadcasts. In 1945, music constituted 34.0% of the total nationwide transmission time. Those were broadcasts from studios, concert halls and operatic theatres, as well as mechanically-reproduced music from gramophone records and magnetic tapes.

Musical activities called for the establishment of independent orchestras, which was not a simple task at that time. There was a shortage of proper studios, musical instruments and accommodations for musicians who were scattered all over Poland. Notwithstanding these difficulties, Polish Radio succeeded in establishing seven orchestras in the years 1945-1949. The largest of them, the Great Symphony Orchestra, was established in Katowice in March, 1945. It gave its first concert on March 25. Initially, the orchestra numbered 18 musicians. After one year, the number of musicians grew to 40, and in 1947 to 54. In 1968, the orchestra numbered 105 musicians.

Polish Radio has set itself the task of popularizing the whole historical achievement of Polish and world literature, and also the modern works. The popularization of literature was assisted by regular evening broadcasts featuring Polish authors, and also by recitation competitions, evenings of poetry, the *Mickiewicz Year*, the *Pushkin Year*, the Slowacki's death anniversary. Polish Radio has also been gathering from the very beginning folk songs and fairy tales.

Not confining itself to broadcasting the works of literature already created, Polish Radio also commissioned well-known authors to write short radio stories. The Polish Radio Theatre, in all transmission networks, broadcast 1600 items a year at the end of the 1960s, including 480 premieres. At that time, the theatre employed dozens of producers and about 1000 actors every year, apart from the directors and producers on the permanent payroll.

In the years 1950-1960, various program forms were developed and streamlined, owing to the introduction of appropriate organizational forms in the management of program production. Hundreds of wholly new broadcasts appeared on the air, notably enriching music programs, show performances and radio plays. Some of them have been turned into serial broadcasts which continue up to the present day: *Five at the Mike, The Matysiak Family*, and *The Jeziorany Village*.

The influence which these broadcasts have had on listeners has exceeded the expectations of their authors. This is evident, for example, in the case of *The Matysiak Family* which for many years keeps drawing thousands of letters from the listeners.

During the early '50s, a number of new radio magazines went on the air. They included *Music and New*, a magazine still broadcast up to the present, which is an unprecedented fact, not only in Poland but also on a world scale.

On November 1, 1957, Polish Radio established an experimental studio in Warsaw, the third such studio in Europe at that time, where electronic devices were used to prepare music illustrations for radio broadcasts and radio theatre, as well as for films. The studio has also been concerned with vanguard music experiments.

A big achievement of the section responsible for political-informative broadcasts was the introduction of the local broadcast for the Warsaw province, entitled *Mazowsze Voice*. It has been transmitted since January 22, 1965.

In the years 1966-1968, many other new political and informative items went on the air. For example, the economic department began broadcasting a *Review of Economic Developments*, the social department introduced a cyclic broadcast entitled *Eight Hours a Day*, concerning the workers problems, the international department introduced a broadcast entitled *People and Continents*, dealing with the most relevant current world problems. Also, new local items were added: the news bulletin, *In Warsaw and Mazowsze* and *The Warsaw Sound Weekly*, which was the Sunday edition of the daily broadcast *On the Warsaw Waveband*. Following the introduction of night broadcasting, the number of editions of the news bulletins also increased considerably.

On February 2, 1966, a 3-hour broadcast entitled *The Afternoon with Youth* went on the air. It was based on the materials supplied by the youth department. Eight specialist departments cooperated in producing the broadcast, which covered a wide range of subjects, such as ethics, modern music, occupational problems of young workers and farmers, lyrical poetry, international problems, favorite hits, fragments of prose writings, interviews with scientists, astronautics, history, the performances of amateur youth bands, discussions on young people's place in life, interviews with favorite actors, problems and conflicts at work.

The *Afternoon with Youth* cycle won many fans from the very outset. In response to the first few broadcasts in that cycle, over 20,000 letters were sent in to Polish Radio. The thousandth broadcast of that cycle went on the air on May 3, 1969. The most popular items broadcast as parts of the *Afternoon with Youth* were the *Radio-stops*: the *Sports Radio-Stop*, the *Nationwide and Worldwide Radio-Stop* and the *Rhythm Studio*.

In the early 1970s, a number of decisions were taken which contributed relevantly to the further development of individual radio broadcasts. Not only had the overall transmission time risen, but also the forms of individual program cycles were enriched and diversified. Those undertakings were crowned by the entry into operation, on July 5, 1971, of the nationwide stereo broadcast, which is transmitted on Radio 3 network. (The first test stereo broadcast was transmitted by the Warsaw station in October, 1967.) In 1972, stereo broadcasts were produced and transmitted also by the broadcasting centers in Gdansk, Krakow, Katowice, Poznan, Szczecin, Warsaw and Wroclaw. In the following year, this group was joined by Lodz. On June 5, 1973, nationwide stereo broadcasts were transmitted three times a week, between 19:00 and 21:30 hours. In 1976, practically all Polish Radio broadcasting centers were capable of broadcasting stereo. The transmission time of these broadcasts also increased.

Another important event in that period was the inauguration of the Radio 4 nationwide broadcasting network, on February 2, 1976. Radio 4 concentrates on educational and music broadcasts, and the most valuable rebroadcasts of literary programs previously transmitted on the other three networks. The concentration of educational broadcasts on Radio 4 left room for other broadcasts in Radio 1 and Radio 2, which before shared the broadcasting of educational, popular-scientific and school programs.

The annual reward of radio critics known as the *Golden Microphone* is held very highly by radio authors. The jury for the award operates at the popular national weekly magazine *Radio and Television*. The award was established in 1969.

DEVELOPMENT OF TELEVISION TECHNOLOGY

Polish television program services came into being as a result of the scientific research performed by a group of Polish specialists in radio technology. The fact is especially noteworthy, as there are few countries in the world, which, like Poland, introduced regular television services on the basis of their own laboratory television equipment.

As already mentioned, Polish research into television had begun before the war. After the war, the research was resumed in 1947 by the State Telecommunication Institute in Warsaw, later the Institute of Communication.

The first public show of television apparatus and picture was held in Warsaw on December 15, 1951, during a radio exhibition which was open until January, 1952. The visitors to the exhibition could watch a program on the screens of TV sets and, at the same time, could observe the work of technicians and performers in a studio separated from the audience by a large glass wall.

At the exhibition, two sets of television studio equipment were shown, one of them an older 441-line (monochrome) system, and the other a newer 625-line system which was eventually adopted for the nationwide service.

On October 25, 1952, on the basis of its own studio facilities and transmitters, the Institute of Communication in Warsaw put into operation an experimental television station. That date marks the beginning of television services in Poland. The next transmissions took place two weeks later, and on December 5 of the same year. On January 3, 1953, a special program for children was telecast.

Regular television services from the Institute of Communication studio started on January 23, 1953. The programs were telecast once a week, on Fridays at 17:00 for between a half hour and an hour. The timing of the program was chosen so as to enable the performers to reconcile their camera appearances with their evening theatre performances.

In November, 1953, the experimental television studio transmitted, for the first time in the history of television in Poland, a full-time production of the theatre play *Window in the Forest*, by Rachmanow and Ryss.

The Friday television broadcasts went on the air for the whole year 1953 and partly into the year 1954. That pioneer period in the development of television in Poland was marked by a good deal of sacrifice, enthusiasm, and fervor.

Those responsibilities for originating TV services required some organizational forms. Therefore, the Bureau for the Expansion of Television (BET) was established in the beginning of 1954 in the Ministry of Communication, with the task of carrying out television investment projects. Until then, Polish Radio was responsible for the program side.

The BET took over, from the Institute of Communication, responsibility for the whole transmission and studio equipment. The institute, from then on, was to deal only with the television projects. The first structure, built by a group of BET specialists, was the Experimental Television Center, (ETC), at 7 Warecki Square. The center, which inaugurated broadcasting on July 23, 1954, had a 240 square meter studio and a transmitter located on the 16th floor of the adjoining Warszawa Hotel. The power output of the image-signal transmitter was 1.2 kW, while that of the sound transmitter was 0.4 kW.

The foundation of the ETC enabled regular television broadcasts which, initially, didn't exceed one hour a week. From November 10, 1954, television programs expanded to two hours a week, and from May 1, 1955, to 100 hours a month. At that time, there were 10,000 receivers in Warsaw.

At the beginning of 1955, the Ministry of Communication initialed and directed the work on preparing the first state document laying down the paths of television development in Poland for the immediate dozen or so years. The document was approved by the presidium of the government and appeared in the form of the resolution No. 21, of February 19, 1955. The resolution envisioned the expansion and modernization of the Experimental Television Center in Warsaw in the years 1955-1956 and the gradual construction of three local television centers.

The wholly modernized Warsaw Television Center (WTC) took up its work on April 31, 1956. That day marked the conclusion of a period of experiments connected with the ETC activities and the beginning of professional television in Poland. The Warsaw Television Center was equipped, at that time, with a 6 kW power output image transmitter and a 2 kW sound transmitter. The transmitters were installed on the 27th and 28th floor of the Palace of Culture and Science. The image and sound transmitters worked through a diplexer, using a common antenna, which was put atop of the spire of the palace, 227 meters above the ground level. The studio was equipped with several superorthicon cameras. The total cubic space of the WTC was 33,000 cubic meters.

Under the government project, the first three local television centers began operation during the years 1956-1957, and all three of them produced and broadcast television programs. The centers were established in Lodz (1956), Poznan (1957), and Katowice (1957), in the most populated areas of the country. Technical limitations made the building of more local centers impossible at that time.

Nobody expected then that the advent of television in Poland would release such great resources of social initiative. Civic committees for the construction of television centers, television fans' societies and other civic groups were springing up spontaneously, with the aim of building a television center in their own locality. The committees brought together politicians, representatives of major factories and industrial associations, the press, radio, etc.

The committees proved unusually vigorous and resilent, a big driving force behind the development of television. They mobilized people and funds. Local factories, enterprises and institutions responded to that large-scale initiative with contributions in the form of money, material and workmanship. Financial contributions from people in town and country also flowed into the committees' bank accounts.

The television construction committees of various levels submitted proposals to the Ministry of Communication and the Polish Radio, concerning the construction of more and more television centers. Under this pressure from public opinion, the state authorities took resolute steps toward providing a legal groundwork for civic initiative and television development ventures. These issues were normalized unanimously by a special resolution of a Council of Ministers, No. 366, of August 31, 1959.

Under the resolution, 50% of the outlays on television projects were provided by civic committees, while the other 50%, in the case of transmitting facilities, by the Ministry of Communication, and, in the case of studio facilities, by Polish Radio.

In practice, civic committees usually financed the construction of a housing structure, while the Ministry of Communication and Polish Radio provided technological equipment. The implementation of the resolution was facilitated in 1959 by the introduction, into domestic production, of the transmitting studio equipment for the requirements of the dynamically developing television network.

In 1960, Polish Television had at its disposal 19 transmitting centers which included 30 transmitters of a total capacity of 272.6 kW. The maximum radiation power of those centers was 2500 kW. The television transmissions covered 63% of the country's territory, or the area inhabited by 78% of the Polish population. All transmitting centers built at that time were housed in modern buildings equipped with antenna masts 330 meters high.

On July 14, 1969, a new transmitting center in Olsztyn was inaugurated. It was equipped with the Polish-designed big-power transmitter and an antenna mast 330 meters high. It covered a large northeastern part of the country, replacing small-power transmitting centers that had served that area before.

In 1970, the number of television stations in Poland grew to 21. They had at their disposal 35 transmitters of a total capacity of 331.2 kW. Five years later, Poland had as many as 47 television stations with 79 transmitters producing a total power output of 556 kW. In 1976, there were 53 transmitting centers equipped with 89 transmitters of a total power output capacity of 656.9 kW. Their maximum radiating power exceeded 6000 kW.

A permanent expansion of radio networks was indispensible to enable the mutual exchanges of programs between television centers. The establishment of the first one thousand kilometers of television links was achieved in Poland in 1959. The link length of 2000 km was reached as soon as 1961. In 1968, the total length of television links was 4200 km.

Today, apart from the numerous local links, there are two international links bisecting the country from the north to the south and from the east to the west. The links are connected to all national television centers. Radio links handle not only the television programs but also a number of other telecommunication services of the Ministry of Communication.

Wide-ranging investments in transmitting facilities were accompanied by huge investments in th development of studio technology. A number of television studio centers were built in the years 1960-1970. The largest of them was erected in Katowice, Krakow and Warsaw. The Warsaw center serves both for radio and television broadcasting and houses the central authorities of the Committee for Radio and Television "Polish Radio and Television."

Poland's largest radio and television project, the new radio-television center in Warsaw, entered into operation on July 18, 1969. The building of the center started in 1962. The Warsaw center is made up of 14 buildings, six of which house television, two house radio, five, technological and service facilities, and one, administration. The total cubic space of the center is 370,735 cubic meters (the space planned is 700,000 cubic meters), the total floor area of all the buildings being 70,219 square meters, of which 14,192 square meters is the area of passageways and transport alleys. The site allocated to the center measures 16 hectares.

The television part of the center is composed of the buildings with the following facilities: studios, central apparatus room, editorial departments, outside-broadcast work shops and stage-setting stores.

The television studio house consists of seven studios of the following areas: 631 square meters, 546 square meters, 400 square meters, 307 square meters, 228 square meters, 92 square meters, and 87 square meters, the total studio area being 2291 square meters. The studios are capable of producing about 29 hours of various programs a week.

The radio section of the radio-television center in Warsaw consits of 14 production studios, among them a large concert studio, two large music studios, a large radio-play studio, three medium and one small music studios and two medium-size radio-play studios.

All the radio-television centers discussed above have been furnished almost completely with technological equipment of Polish make. This is true also of such complicated electronic devices as videotape recording equipment which has been produced for over 10 years now by the Radio and Television Research Development Center in Warsaw. The video devices are used to record and reproduce both monochrome and color images. They match the parameters of even their best-known counterparts made by world-renowned firms.

The television (and radiophony) production capacities are expanded additionally by permanent outside broadcasting points in cultural and sports facilities. The Radio and Television Committee "Polish Radio and Television" invests considerable outlays for this purpose. Practically all major public structures built over the past 30 years have been equipped with permanent radio-television installations and some of them have complete radio and television equipment.

Since 1970, other big projects in the field of television studio technology have been undertaken. The most noteworthy are the TV film production enterprise and several big regional television studio centers. The centers are under construction in Poznan and Gdansk, among other cities.

In the period past 1970, a large-scale modernization of installations in the existing television centers also took place. This was connected with the introduction of the second central television network and of regular color services, based on a SECAM system. The color services are transmitted both by the first and second television network.

An important event in that period was the putting into operation on Poland's first satellite communication station in Psary. The station is used to transmit and receive black-and-white and color television signals together with the accompanying sound track, as well as to transmit simultaneously a number of telephone calls and other signals sent out by various telecommunication services. The Psary station is working in the Interspoutnik satellite communication system. It relayed its first program on July 17, 1974. Poland officially joined the space communication network on November 17, 1974.

As a part of a wide-ranging government project, now implemented for a few years, Poland is aiming to start up production of color TV sets. A large color picture tube and a number of other electronic enterprises, which will be making components for the color sets, are now under construction.

DEVELOPMENT OF TELEVISION PROGRAMMING

As already mentioned, regular television services began in Poland in 1953. In 1955, a total of 214 hours of programs were telecast.

The turning point in television development was the inauguration of the Warsaw Television Center (WOT) on April 30, 1956, and of the three regional centers. From that time, the television programs wre transmitted five times a week, four to five hours a day. By the end of 1956, WOT broadcast a total of 647 hours of programs, while local centers transmitted a total of 110 hours.

On July 1, 1958, a separate team for television programming was set up within the Committee for Radiophony "Polish Radio and Television." From February 1, 1961, television programs were transmitted seven times a week: on Saturday six to seven hours and Sundays, 11 hours.

The overall shape of the nationally-transmitted program schedule became evident in those years. Formally, the framework schedule was formed on September 1, 1958. But, it was not until the end of 1960 that the Warsaw-telecast programs could be received by all local centers. The first entry in the nationwide program was officially inaugurated from Warsaw at that time also.

However, not all of the television centers established at that time had the capability to transmit their programs nationwide. There was a shortage of relay stations to beam programs in varying directions. It was not until 1962 that the possibilities concerning such relays had been fully utilized. Therefore, nationwide transmissions before 1961 were programs originating in Warsaw, although there had already been many programs produced by local centers.

In 1967, the individual centers transmitted on the average of over 3000 hours of programs. The Katozice center provided the longest transmission time.

The situation has been unchanged since 1961. This is explained by the fact that the Katowice center was the first to start broadcasting before noon for viewers working on the second shift. This broadcast is a repetition of the evening programs from the day before.

The organization of television programs in the '50s was similar to that of radio programs. The television unit was divided into four chief editorial departments: current topics and information, popular science, artistic, and children and youth. The chief artistic department was responsible for music and theatre.

The television theatre, since the beginning of its existence, has represented a high artistic level and many of its productions can certainly be described as great theatre events.

The tradition of the Monday theatre is still very much alive today. On Monday, television presents the most valuable items of Polish and foreign dramaturgy, the best stage managers, producers and actors. No less popular with the viewers is the Thursday's *Thriller Theatre*, though, naturally, the reasons and motives behind that popularity are different.

As regards the TV *Thriller Theatre*, the most popular with viewers had been the series entitled *More Than Life Is At Stake*, which was also presented later by many other television networks in the world. The first part of the series was telecast on January 28, 1965. At the end of 1965, a decision was reached to make the series into a film. The first screenplay was approved for production in November, 1966, whle the first of the 14 installments of the series was screened by television on October 10, 1968.

The tremendous interest in TV theatre gave rise to the fine tradition of the *Golden Mask* and *Silver Mask*, the awards conferred by the viewers on the most popular actors and performers in a given year. The poll, which attracts over 100,000 people every year, is run by the editors of the popular Warsaw evening Daily *Express Wieczorny*. The Golden and Silver Masks were awarded for the first time in 1961. The first edition of the poll recorded the participation of nearly 114,000 people from all over Poland. In 1968, a total of 180,000 poll entries were made.

Another important component of the television program schedule are educational and popular-science broadcasts. In this field, Polish television has become one of the basic links in the whole system of civic education and upbringing. The telecast educational programs were introduced experimentally into schools in 1959.

Television took up regular transmission of educational programs on March 2, 1961. Initially, the programs were transmitted three times a week and lasted half an hour a day. The television educational broadcasts were especially enthusiastically welcomed by schools outside large cities. By the end of 1962, educational programs on television had been viewed in over 3000 schools of all kinds and types. In 1964, they were viewed in 7500 schools and in nearly 10,000 schools in 1968.

In the following years the school programs were enriched by new cycles and subjects and applied to a greater number of school grades. The programs added a precious diversification to the forms of didactic and educational work by introducing ever-new audiovisual aids, particularly in physics, chemistry, biology, and geography.

However, the real highlight was the introduction of technological courses on TV in 1966, in consultation with the Ministry of Higher Education. The courses are designed for the higher-level studies of technology for working people.

The opening lecture in the television technological courses was broadcast on September 4, 1966. The first-year technological course and the total number of viewers of both was estimated at 12,000.

It is worth noting that students of TV technological courses have the right to take exams qualifying them for the second year of studies in the Institutes of Technology. Poland and UNESCO have signed an agreement under which Poland would make its experiences with the TV technological courses available to that organization.

Apart from the programs mentioned above, educational broadcasts also have included foreign-language courses and agricultural courses, since October 7, 1961.

The television university is different from the aforegoing programs for organized viewers, as it is addressed to all. The TV university programs are of a popular scientific character. Their aim is to enrich knowledge, shape the outlook on life and help viewers understand current scientific and technological developments. The TV university was made up of cyclic programs, such as *Eureka*, the *Technical Progress Magazine*, the *Medical Magazine*, the popular *Meet Nature* series devoted to architecture, entitled *With a Drawing Pen and a Crayon*.

A large group of programs are shown along with artistic broadcasts for children and youth. The TV *Threatre for Children* has staged many shows and plays concerning childrens' everyday life in present times, like the *Bedtime Stories*, introduced on October 3, 1962, where a couple of kids named Jaced and Agatka and a hero of one of the first such programs, *Teddy Bear from the Window Square*, were the favorites.

In the 1950s, programs concerning current affairs were a sideline in the television program schedule. Their real development dates back to 1960. One of the first informative programs was the *Television News*, the first edition of which went on the air on January 1, 1958. The news became a fixture on January 29, 1958. At present there are three extensive daily editions of the TV news: the afternoon, evening and night one.

The local news bulletin for Warsaw and Mazowsze went on the air on November 22, 1958, and later was followed by a second program of the same kind, their titles being, respectively, *The TV Warsaw Courrier* and

The TV Mazowsze Courrier. The two bulletins bring current information on the most important events in the Warsaw and Mazowsze regions. On February 24, 1960, the first edition of *The Periscope* was transmitted on the nationwide network. In a short time, that program has become a leading item. Later, it was transferred into a TV news magazine *The Monitor.*

The information department also broadcasts programs: *Sports News, Medals and Details, Physical Training of Our Children,* as well as live sport telecasts, commemorative films, edited films and live reportages. The same department also initiates programs designed to stir public opinion: *The Green Light Calendar Cards, Somewhere in Poland, Concise Encyclopedia, Culture Chronicle* and others.

Programs on current affairs concern social life, politics and economy. One of the first programs of the social cycle was *Tele-Echoes,* on the air from 1957 till the present. It consists of interviews with politicians, economists, cultural workers, artists and of performances by singers, musicians, recitors and dancers. Among the other social programs were the following: *Reflections,* on the social life; *The Fourth Shift,* on workers' problems; *Without Appeal,* on social conflicts and customs; and many other programs produced by local centers, such as *On the Orda River and the Baltic Sea, Outstanding Matters, Trials, Hotline, The Breeze,* or *Away at Sea.* These programs deal with national and regional economic problems.

The rural department has also made a large contribution to the program schedule, with such highlights as *The Changes, What is Best, Good Farmers' Club.* The military department has produced *The Practice Range* and *The Azimuth,* and the international department, *People and Events, The Tournament of Towns.*

A permanent component of informative and public-affairs programs are the commentaries and dispatches sent in from Polish radio and television correspondents abroad.

Apart from the nationwide broadcasts, TV viewers can also see Polish and foreign programs transmitted through the Intervision network. The first Intervision program was transmitted for Polish viewers in 1960. On January 28 of that year the OIRT established Intervision, a system of program exchanges. The first program offered by the Eurovision system was transmitted to Poland on July 6, 1954. It was the *Narcissus Festival* from Montreux.

In 1970, Polish television received from abroad a total of 299 programs of 310 hours total broadcasting time, including 243 hours of Intervision programs. In the same year, Poland transmitted abroad 49 programs totalling 117 hours. Polish television is currently cooperating with nearly 90 television organizations in various countries.

All the fields of program activity discussed here use about 70-75 of the transmission time. The remaining 20-25% are films. The statistical Pole watches more films on TV within one month than he sees in the cinema within one year.

In 1969, television programs concentrated on the 25th anniversary of the People's Poland, reviewing postwar accomplishments in every field, as well as the postwar literature. For example, the TV theatre included in its repertory, alongside original new script written especially for the occasion, works by prominent Polish writers and playwrights in new production, as well as the repetitions of the most outstanding plays such as *The Germans*, and *The First Day of Freedom* by Leon Kruczkowski.

Since 1970, relevant changes have been introduced into the television program schedule, both with respect to the quantity and quality of programs. Important determinants of the development were the introduction of regular services by the second television network, and the introduction of regular services in color.

The first color program was transmitted from the studio on March 16, 1971. At first, color programs amounted to six hours a week, but they totalled 16 hours a week in the second year.

In those years, many new items were added to the program schedule. Educational and popular-scientific programs have been considerably expanded, by the second TV network particularly. The new cycles in this field included *Computers for Everybody, The World of Physics, Mathematical Stories, Organization of Work and Management, Chemistry of Large Synthesis, Mysteries of Electronics,* and *Man and Egonomy*.

A notable part of the second network transmission time is taken up by cultural items. In the reviewed period, Polish TV introduced also the Teacher's Radio and Television University (NURT), with the purpose of improving the professional qualifications of the teachers. The NURT programs are currently viewed by over 100,000 teachers who have committed themselves to raising their knowledge, not counting the sympathizers of NURT lectures, in psychology, pedagogics, mathematics, etc. The majority of lectures are printed in a special supplement to the bi-weekly *Education and Upbringing*. A large circle of viewers is also attracted to the television agricultural school. In 1976, the school provided aid to 30,000 students.

The Radio and Television Committee awards annual prizes to writers, journalists, actors, announcers and broadcasters for their contribution to the creation of radio and television programs. The first prize of this kind was awarded in 1957, so the year 1977 marked 20 years since the prize was established.

PROSPECTS FOR THE FUTURE DEVELOPMENT OF RADIO AND TELEVISION

The Polish radio and television enterprise, like many other such organizations around the world, is engaged in systematic research into the further development of technology and programs. The studies lead to drafting short- and long-term tasks. Short-term tasks are usually scheduled for implementation within a period of five years. Such tasks, particularly if they concern the development of radio and television technology, point out the scope of the new investment projects planned and the material and financial means needed for them.

Needless to say, the development of technology has stimulated and will still stimulate largely the radio and television program capacities. On account of this fact, efforts have been made to expand further the existing studio and transmitting facilities, both in Warsaw and in individual regions.

The largest producing center among the studios will still be the central radio-television broadcasting center in Warsaw, which will be substantially expanded in the years immediately ahead. Also, a number of radio and television studios and transmitters in regional cities will be considerably expanded, notably in Poznan, Gdansk and Krakow. The largest transmitting long-wave center will still be the 2000 kW transmitting center in Gabin, and the largest medium-wave center will be the 1500 kW station in Katowice.

Today Polish radio runs four nationwide program services. The fifth radio service will soon begin operation. It will be largely made up of programs helping parents in their children's upbringing. The introduction of this service is sure to be a great success for Polish radio broadcasting. The service is likely to become a model and a source of experiences for other broadcasting services in the world.

Polish television runs two nationwide services, part of both of them in color. Apart from the nationwide programs, local enters transmit to a wide extent their own radio and television programs. In the years immediately ahead, a share of stereophonic radio broadcasts and color TV programs will keep growing.

The structure of radio and television programs in the future will doubtlessly be affected by the changes in the structure of Polish society. Even today a majority of the country's population inhabits urban agglomerations. The working time is being shortened. These factors will increase the amount of free time gradually; consequently, the search for new forms of psychic relaxation and physical recreation will grow. It is assumed that radio and television programs will largely help satisfy such needs, if only they are appropriate. There is also a widespread belief that more and more programs will be produced, and there will be participation of the public, that is, in direct contact with the recipients of radio and television programs, in lighter fare as well as in social-political and artistic broadcasts.

Radio and television also have a growing part to play in the educational area. In view of the need for the continuing education of society, radio and television will represent an increasingly important link in the whole system of national education. These functions lead to conventions and methodological-didactic tasks that run far beyond the scope of the merely verbal transfer of knowledge. The up-to-date achievements of radio and television in this field have been reviewed before in this work. It should be presumed that these achievements will still be in the focus of interest of various international organizations, especially UNESCO.

A more ambitious schedule of radio programs in quadrophony remains an outstanding issue. As it has already been pointed out, Polish radio was one of the first broadcasters in the world to take up regular quadrophonic transmission, though, for the time being, on a very limited scale.

Research is also underway into the methods for transmitting additional information by means of the television signal. Such additional pieces of information include, among other things:

● captions designed for people with impaired hearing, subtitles for foreign-language films, subtitles in two languages, etc.

● periodic screening of the titles of programs on view, and the number of the TV network

● program schedules enabling the viewers, while they are away from home, to record selected programs on videotapes

● television newsreel to be received by a viewer on his TV screen instead of a regular program.

● news scraps screened against the background of an ongoing program

● transmission of the second and even more sound signals.

Apart from that, the TV set can be utilized as a computer terminal connected with a computer center by a telephone line or a special wire line.

The transmission of two sound signals to accompany the image is of particular importance to TV viewers. Many viewers prefer listening to the original film dialogues instead of having to listen to a Polish reader. In such a case, the original sound signal could be transmitted by an additional sound channel. The viewers would then have the possibility to choose between the language versions transmitted concurrently, one of which could be selected by pressing a respective button or actuating a special adapter. The possibility of introducing the stereo transmission of sound into television is also being examined.

Temporarily, Polish television has been testing the viability of using radio receivers to receive additional channels of mono or stereo sound. The mono or stereo sound accompanying some stereo programs on TV is transmitted simultaneously by a television transmitter and a VHF radio transmitter. The first test of this kind, with the TV image accompanied by the stereo sound, was carried out on April 18, 1977, on the first TV network and Radio 4. A similar program was transmitted on October 16, 1977. This method is also expected to be employed to transmit original-language film versions.

Almost all equipment now in use in Polish radio and television centers, like in other broadcasting centers of the world, is based on the analog technique, that is, the technique in which electric signals oscillate, continuously reflecting the corresponding acoustic or optical oscillations. However, radio broadcasting and television are headed inevitably toward the wider use of the digital technique. This has already become a fact in some branches of teletransmission, and the virtues of time systems with code-impulse modulation support the expectations that this system will rapidly supplant the traditional one. The introduction of the digital technique will mark a successive revolution in radio and television studio technology. The high quality and stability of this technique will make a number of today's parameters irrelevant. A full self-diagnostics and self-correction of failures in installations will become a fact.

Many research centers in the world are currently engaged in studies on the application of the digital technique in radio and television broadcasting. This technique is also taken into account in the research and design work carried on by the Polish scientific-research centers, as they are now constructing new, prospective generations of radio and television studio equipment.

A large-scale work is also in progress in connection with the automatization of technological processes in radio and television studio centers, especially with respect to those stages of the continuity processes that are related to the continuity of radio and television programs. Polish radio broadcasting houses are already applying some technological solutions of this kind.

The first computer systems capable of retrieving information from the archives have also been introduced. The electronic calculating technique is also made use of by the finance services of Polish radio and television. Such systems will gradually encompass ever-new components of the complex, dynamically developing radio and television services in Poland.

In the distant future, now hard to define, one should expect the advent of a very high quality of television. Most probably, it will mark the transition stage prior to the introduction of spatial television which will reproduce also the depth of the transmitted scenes. Since it will be both color and spatial, it could be called integral television. The color-spatial television technique will fulfill all conditions of the faithful transmission of the scenes actually existing in the studio.

Other directions in the prospective development of radio and television services are shaping up in connection of the expansion of satellite transmission. The realization of the idea of satellite television will require solutions to many difficulties of a technical nature, but not only those; it will create a range of extratechnical problems which will call for extensive—and not at all easy to achieve—international agreements.

One could give more examples of such prospective technologies. Each of them is in its way attractive from the viewpoint of radio and television broadcasting. The future will tell which of them turns out to be the best one at a given stage of development. Anyhow, the variety and high qualities of the prospective techniques warrant the expectations that they will become a source of new informative and artistic possibilities, new categories of human experience, and that they will enrich and inspire the new, now unknown forms and types of programs. Moreover, those techniques that will facilitate rapid response to proposals and inquiries forwarded by radio and television viewers will increase the direct contact with them. This is, after all, the main objective toward which all duty-conscious radio and television boradcasters are heading. However, it is worth stressing though it can be taken as a truism, that both today and in the future, the techniques will never substitute for the creative inventiveness of the producers and performers of the programs.

Chapter 3
Broadcasting in Russia

Vladimir Yaroshenko
Ph.D. in Historical Sciences,
Lecturer at the Broadcasting and Television Department,
Moscow State University

The trajectory, origins and development of broadcasting in Russia were closely interwoven with the emergence and making of the Soviet Socialist State. Ever since the first day of the Soviet Republic, in 1917, radio has served it.

BEFORE THE REVOLUTION

In Russia, radio emerged independently of its appearance in other countries. In 1859, the Russian physics researcher, Alexander Popov, presented a device capable of displaying "electromagnetic properties." On May 7, 1895, at the meeting of the physics department in the Russian Physics and Chemistry Society held at the University of Petersburg, he came forward with a public report and demonstrated the first radio receiver in the world. In the Soviet Union, May 7 is considered Radio Day.

In the year 1900, radio had already been used to save the lives of fishermen that found themselves on ice broken by gales in the Finnish gulf of the Baltic Sea. In February, 1904, at the third all-Russian Congress of Electrical Engineering, Alexander Popov, equally for the first time in the world, demonstrated the broadcasting of live human speech at a distance, without cable. However, until 1918 the only form to be actually used on a large scale was radiotelegraphy.

The development of radio in Russia before the revolution went along lines followed by the development of all industry during the Tsarist regime: in 1907, radio was sold to "Marcon" and subsequently joined by "Siemens" and "Telefunken." A Russian national radio broadcasting company was created later, in 1910, through Popov's efforts.

AFTER THE REVOLUTION

In the first years of the Soviet Republic the rapid development of radio broadcasting can doubtlessly be ascribed to the founder of the Soviet

State, Vladimir Lenin. In those years, hard for the young Soviet Republic—the Civil War, the intervention of 14 capitalistic states, the famine, the ruin, the want of goods and equipment—Lenin found the time, means and people for the development of radio broadcasting. The importance of broadcasting as a means of enlightening the large masses in Russia was unquestionable at that time, since the overwhelming majority of the population was illiterate. Lenin underlined not only the high efficiency of broadcasting, but also its ability to travel rapidly any distances. This policy also offset certain difficulties that faced the republic: lack of paper, poor connections and transportation, especially stringent in a country as widespread as Russia.

The construction, soon after the revolution, of a radiolab in Nijniy Novgorod (now the town Gorky), whose mission was the elaboration of a means of communication, enjoyed the attention of Lenin. The researchers in this lab were being given nourishment and clothing. Here is one of Lenin's statements about broadcasting: "A thing of tremendous importance since it is a newspaper with no paper or wires, for by means of a transmitter and a receiver (adapted by B. Bruevitch, so that we will be able to get hundreds of them) all Russia will hear the paper read out in Moscow."

The first telegrams of Soviet Russia addressed "to the citizens of Russia," the decrees "on peace," "on the world," "on the creation of the Soviet Government" were heard not only in our country, but also abroad in Germany, France, Austria.

GROWTH AND CONTRIBUTION OF BROADCASTING

On February 27, 1919, the radiolab in Nijniy Novgorod started broadcasting live speeches. These were experimental telephoned radio programs. Sometime later Moscow started broadcasting news bulletins. During the summer of 1921, in six of the main squares in Moscow loudspeakers were installed; they were relaying the news section of the radio journal. Daily (except on rainy days) from 9 to 11 p.m. programs were aired, including articles from the paper, lessons, reports. The inhabitants of the capital were constantly aware of what was new in current affairs at home and abroad, as well as of the weather forecasts.

On August 21, 1922, a central radiotelegraphy station started operating in Moscow. On November 7, 1922, for the 5th anniversary of the Great October Socialist Revolution, a special program was aired, dedicated to the event.

In May, 1922, the station at the Nijniy Novgorod lab broadcasted experimental programs of musical concerts. The broadcastings could be heard within a range of 2000 miles. The first concerts consisted of works by Tchaikovsky, Borodino, Rimski-Korsakov and other classics. Radio broadcastings did not overlook the heritage of the world musical culture and aired both classical and folkloric national music, not only Russian but also that of the peoples throughout the country.

Beginning in 1922, the broadcasts were transmitted if not regularly at least more or less systematically. Programs included both news items and artistic and literary shows.

After surviving the civil war period, as well as that of the revolutionary reconstruction of the political, economical, social and ideological life of the country, radio broadcasting in the Soviet Union could in no way fail to be in the most advanced ranks of the political struggle. The spreading out, due to the radio, of correct information about the surrounded and beseiged republic was met with adversity by the western countries. Already at the time, and not in 1934, was stated by many historians, brodcasts of the Soviet Republic started being jammed. Lenin said in 1918: we have reasons to believe that our radiotelegram on our victory over Kerensky was intercepted by Austrian radio operators and broadcasted. The Germans sent forth counterwaves to jam it.

It was up to radio broadcasting in the Soviet Union to play an important part in eliminating the century-old backwardness and in carrying out the tasks facing the republic: industrialization, increasing work productivity, solving problems conneeted with the development of agriculture, organizing the collective farms.

On July 28, 1924, the Council of the Popular Commissars (the government) issued a resolution on the local receptions of radio stations. This resolution is known as the freedom-of-broadcasting law. This law conceded to private organizations, as well as individual collectivities, the right to establish, utilize and profit by radio stations. In that same year the Society of Soviet Radio fans called "The Radio Fan Society" was created. It made an important contribution to the development of radio broadcasting in the 20th century.

The professional unions played a great part in the development of radio broadcasting. The Moscow council of the professional unions founded a special committee whose task was to help the development of radio. This same council sponsored the creation of its own radio broadcasting station, which was one of the most powerful in the country.

In 1924, the following magazines were issued: *Radiofans*, published by the Radiofans Society, and *Friend of Radio* in Leningrad. The technical radio magazine of popular science, *Radio for Everybody*, was issued, starting in 1925. In that same year a daily started being published, *Radio News*, which was the organ of a society called *Radio Broadcasting*. In 1924, 10 radio stations were functioning.

Two more powerful stations were transmitting from Moscow: one was working for the "Komintern" (Central) and one in Sokolniky, named after Alexander Popov, servicing the professional unions. Besides those, there were stations functioning in Leningrad, Kiev, Minsk, Rostov, Tashkent. New stations were built in other smaller towns. The Komintern radio station was directing its programs to all the population. The station in Sokolniky was broadcasting mainly for workers.

PROGRAMMING DEVELOPMENTS

In the beginning, radio borrowed many genres and forms used in the press to the point of calling individual programs, journals. So two radio journals were aired on the waves: the first was a daily one for workers and the second for farmers was aired three to four times per week.

Starting in 1924, stable, regular broadcasting was initiated. The Komintern central radio station was airing programs, universal as far as topics go, various as far as genres and forms, and intended for the masses of listeners in the country.

Starting in November 23, 1924, the main current events were worked into a regular program: *The Journals of the Russian Radio Telegraphic Agency*. This program could not merely repeat the text of the stories and other news received. It was the principal *other* form of mass media, one that took into consideration the specific of radio broadcasting—namely the fact that in order for a material to be comprehended by hearing it, it has to be expressed in a different language. There were attempts made at differentiating between programs according to different categories of listeners' social, professional and demographic characteristics.

Radio broadcasting became one of the major means of education and instruction especially for the rural population whose great majority were illiterate. Under these circumstances, equipping the whole country with radio installations was viewed as a most important step at that stage, even over the creation of libraries. For that it was imperious to channel the programs in such a direction as to carry them out in such a way as to be understood and judged interesting by the farmers. That is how special radio programs for the rural population were initiated.

In rural areas there was not enough material basis for a speedy eradication of illiteracy; there were few schools, even fewer clubs; there was no paper; and finally, there were no experienced teachers capable of not only teaching reading and writing but also of conveying universal information in various scientific fields. Thanks to its peculiarity, radio made it possible to use the best teachers, scientists, artists for the cultural renaissance in the rural area.

The first issue of the radio journal for farmers, broadcasted by the central station, included, for instance, material such as: information about the country's life in *What's new in the Soviet Republic* (about the election of organs of power, the preparation for seeding, the processing of grains, the creation of farmer's unions, the sponsorship of villages by cities); world news in *What's New Abroad*; a section called *What's New in Moscow* (comments on the main political events); a series of communications about what the central ministries do for the villages, a section called *Medical Advice for Farmers*; an informative legal section (information and short consultation about loan of seeds, taxes, state insurances); a section called *Useful Books for Farmers* (literature for agriculture).

Radio broadcasting was vital also to the education and formation of the population of the national republics, especially in the far-out parts of the country, where there were many such nationalities not as yet familiar with written literature or with written language for that matter. Radio, first of all, created a normative common language, then spread among the national minorities their own national culture and familiarized the listeners with their cultural heritage.

One of the main principles followed by state broadcasting ever since its first years was that of having local programs in national languages. They carried to the people who didn't or couldn't as yet read, information related to medical care, sanitation, and recommendations as to how to improve their household.

In 1927, on the occasion of a jubilee session of the central executive committee of the USSR, dedicated to the 10th anniversary of the October Revolution, the manifests endorsed by the session mentioned as a remarkable achievement that radios and tractors were the first accomplishments in the villages. In the year 1927, one of the most powerful transmitters in Europe for that time (40 kW) started functioning in Moscow. This significantly enlarged the range of operation for the central radio station. The tractor and the radio were two aspects of social revolution in the rural areas: the tractor marked economic aspects, radio the cultural aspects.

In the years 1925-1927, separate departments were created in the central station, and local broadcasting became more and more seriously involved with critical programs. These departments directed their programs to various age and professional groups of people. In 1925, within the framework of central broadcasting the news program *Radio Pioneer* was set up for first graders and *Radio October* for high school children. From that year to this day different programs have been aired for children of various ages.

Starting with the year 1925, reports of events were broadcasted on location (at first without a commentary). These were the predecessors of live radio reports, which in 1927-28 represented very widespread genres.

Documentary broadcasting, with its realism, its possibility of encouraging listeners to participate in what is going on, made radio highly popular. On May 9, 1925, the meeting of the 12th All-Russian Soviets Conference was aired. More than 200,000 people listened to it. In the same year, 1925, the Moscow radio station broadcasted from Bukara the debate of the Uzbekistan Communist Party.

In the capitalistic countries around the Soviet Republic, where a very significant section of the population was familiar with the Russian language, it was prohibited to tune to such programs. In addition to that, the radio stations of many countries initiated what was called at the time "cat's concerts"; that is, they aired artistic programs to jam the Soviet broadcasting.

On May 1, 1928, a report was aired from the Red Square in Moscow by correspondents located in a special stand there. Starting with the same year, 1928, one could hear on the air at morning and at midnight the striking of the clock in the Spaski Tower in Kremlin. The microphone was connected on location for a couple of seconds so that one could hear not only the clock striking, but also the sounds in Red Square. The hum of the Red Square and the strikes of the Kremlin Tower clock can be heard on the first program to this very day. Starting in 1978, this signal began being aired by Radio Moscow's World Special Programs in English.

Toward the end of the '20s, radio broadcasting was virtually directing the national republic; programs on location were largely used; programs were developed by the participants of each and every country. They were not a mechanical repetition of or even molded after the program of the central radio station; instead, taking into consideration national, geographical and economical peculiarities, they each helped in their own way the development of their respective countries.

The rapid development of radio broadcasting did not have a negative influence on the other means of mass information; on the contrary, it strongly stimulated new questioners among the listeners, broadened the range of their interests, increased the need for knowledge. It was an effective device for developing the national culture, and overcoming the century-old cultural isolation.

In 1926, the radio station in Baku (the capital of Soviet Azerbaidzan) started broadcasting in the Azerbaidzan, Russian and Armenian languages. In January, 1927, in Tashkent (the capital of Uzbekistan) a radio program was created on the basis of the paper *The Pravda of the East*. On March 27, 1927, the Tashkent radio station, the most powerful in the mideast, started airing programs in the Uzebekistan, Turkmenian, and Kazakistan. In May of that same year, live artistic programs were aired from Leningrad, Kiev, Harkov, and Tbilisy. In 1928, there were 65 radio stations functioning already throughout the country.

The use of radio equipment by stations in the far-away corners of the country resulted in funny incidents. When the young people in the mountain village of Gudermsk were installing an antenna, radio receivers and loudspeakers, they were met with laughter. "It's easier for a donkey to chirp like a nightingale than it is for your contraption to speak in a human voice," said the old people. But one evening the loudspeaker started relaying music and songs and it was a true treat for the whole village.

The development of radio broadcasting stirred up strong arguments among the workers in the field of arts compatible with broadcasting about the possibility of the existence of a very radio art, about the degree of specificity of this new channel of mass information. These debates were carried on in the informational political programs as well as in literary and dramatic ones. They gave birth to terms such as "radio phonics" and "local live broadcasting effects."

After the establishment of radio broadcasting, the next step was creating high-quality programs and then the process of realizing the peculiarities of radio reporting and writing for the radio. This was by no means an easy process; but, it was a natural one, since the obvious step, after the mechanical transfer to radio of newspaper journalism methods and methods of music, variety performance, concert, dramatic and operatic theatre, was the emergence of specific genres of broadcasting and writing for the radio.

On the occasion of the first International Artists Convention of the Soviet Union, which took place in 1928 in Germany, Soviet radio gave to other stations an example of experimental endeavors used to popularize Russian and Soviet literature.

Between 1927 and 1929, radio broadcasting for foreign countries started being developed. Until 1929, radio programs were sporadic. In

October, 1929, a department for broadcasting in foreign languages was created within the Soviet broadcasting system. The first programs in German were broadcast during the day. They were aired twice daily and their length ranged from 30 to 90 minutes.

On November 7, 1929, a program in French was initiated. It was dedicated to the celebration of the October Revolution anniversary. These programs were aired on Sundays and lasted an hour. Toward the end of 1929, programs in English were initiated.

It should be mentioned that ever since the first years of Soviet broadcasting the main task, as far as programs directed to foreign countries were concerned, was to throw light on the life in the U.S.S.R., to present the achievements of socialist reconstruction, to impact our cultural treasures to foreign listeners. The programs were using TASS materials, information from Soviet papers, and especially from the foreign language papers issued in Moscow. Since that time to the present day, more than half of all the texts used for Moscow radio programs directed abroad focus on internal Soviet issues; a significant part consists of musical and literary materials.

Literary reviews and musical concerts have been included in programs ever since 1930. Before that such programs were broadcasted live, not recorded. This was possible, basically, with programs for such European countries where the difference in time allowed for artists in Moscow to perform at convenient times of the day.

Literary and artistic programs for foreign countries used works of Russian and Soviet classics, and, equally, dramatizations of artistic works. Well-known musical ensembles performed in front of the microphone: the Krasnoznamensk singing and dancing ensemble of the Red Army, conducted by Alexsandrova, the Academic Choir conducted by Sveshnikova.

In 1933, the Moscow radio station was airing programs for foreign countries in German, English, French, Hungarian, Spanish, Italian, Czech and Swedish languages. Besides that, some local stations also broadcasted for foreign countries; Kiev, for instance, aired programs for Poland. A remarkable response was elicited by the programs for Spain at that period, for as many as 600 letters a month arrived. In the year prior to the Spanish peoples antiNazi war, Moscow radio acquired a wide popularity in that country.

Toward the end of the '30s, there were programs in 10 foreign languages, amounting to 20 hours per week: five in German; three in English; 3.5 in Italian; 1.5 in Czech; and 1.5 in Spanish and French; one each in Dutch, Swedish, Hungarian and Chinese. Spontaneously, before the Great Patriotic War (1941-1945) programs were initiated in the Bulgarian, Finnish, Romanian and Serbian languages.

At the beginning of the war, the Soviet Union had 6975.2 thousand radio broadcasting stations. Of them, 1122.5 thousand were wireless stations and 5852.5 thousand were telegraphic relaying stations.

Radio listening at that time was often done in groups; each radio station was then servicing several dozens of people. That made a listening audience of several million. By 1941, a fairly solid radio system was created. The centeral system (the central station serving the Komintern) was airing programs for the whole country. Besides, there were stations functioning at republic, regional, country, and local levels.

In 1936, the central radio committee launched into operation five broadcasting networks: central European, mideastern, west Siberian, east Siberian and far eastern; that is, the central broadcasting station was airing five variants of the initial programs, which took into consideration geographical as well as time differences. Before the broadcasting and the preparation of the program, radio reporters took into account the national aspects, language and other specifics of the representative areas. In that period the central shortwave station, R.V. 96, started functioning with a power of 120 kW.

The main functional principle of Soviet broadcasting is the combination of central and local broadcasting. Each local station with its radio committee, has, of course, its own network. However, before the creation of local networks, these committees took into consideration the necessity of including into the local programs the broadcasts of the central station in order to keep the population more efficiently in touch with Soviet and world news, as well as to provide listeners with high-quality artistic programs.

As opposed to many other countries, whose territory by comparison is small, Soviet broadcasting, the first in the world in area covered, used not only medium and long waves but also short waves. Programs were aired on short wavelengths by the central radio station, the Moscow station, as well as by radio stations of those republics and provinces whose territory was very big: Kazakhstan, Ukraine, the countries of Siberia, and the Far West. Starting in May 20, 1940, the central network had programs on three independent stations, the first, second, and third.

In the years prior to the war, radio broadcasting had an important part to play in the political, economical and the cultural education of the workers and farmers. There were special programs for the workers in industries such as coal, metallurgy, and textile; for railway workers, for collective farmers, teachers, doctors.

The year 1939 witnessed a broadening of broadcasting for workers in the field of agriculture. The activity of this section was carried out with a view to helping the farmers master new methods in agriculture, the new techniques, to share their experience, to tell about the achievements of individual farmers, collective farms, and republics. Experts and leading figures in agriculture could be heard on the air.

The reorganization of the educational system took place at about that same time. At the end of the '30s a broadcasting university was opened within the framework of the radio committee. This was progress compared to the educational system in existence until that time. The programs of this university were intended for listeners, the majority of which were and are of this 7-year school.

Much attention was paid during these programs to international affairs, to the analysis of economical, social, historical processes undergone by the other countries. As the international situation became more and more entangled at the beginning of World War II, one could hear on the air talks, editorials, commentaries, and reports that exposed the aggressive acts of Fascist Germany and Italy. There was an increase in the bulk of broadcasting material that underlined the necessity to strenghten the power of defensive countries and what to do to carry out this task.

During World War II, between 1941 and 1945, radio played an exceptionally important role. It became an unusually important instrument of informing, mobilizing, strengthening the population, a principle means of fighting the enemy. One cannot imagine the history of that war without the contribution made to it by radio broadcasting.

During the war, the central Moscow station switched its programs onto short wavelengths, using comparatively small shortwave broadcasts whose power ranged between 3.5 and 10 kW. Instead of the three programs, only one was aired. It had, as a whole, live reports. Many radio stations in the western and central countries, Bielorussia, Ukraine, Russia, were moved into eastern countries. Many stations in Moscow and Leningrad were also dismantled and moved to other parts of the country. The volume of broadcasting in the U.S.S.R. during the war did not drop, however; on the contrary, it rose from 253.2 thousand hours in 1940 to 267.4 thousand hours in 1943 and 325.8 thousand hours in 1945.

A quite powerful station for that time, 1200 kW, was built in order to maintain contact with the partisans and to service those territories temporarily occupied by the Nazis. It was built in 1942-1943 in a short time, and the equipment for it was manufactured in the besieged Leningrad. Ignoring the terrible working conditions, the famine, the intellectuals and workers of Leningrad built a radio station and transported it to its location through the battle line.

In Leningrad itself, a powerful broadcasting station was created and it was assigned an historic task: to play the part of the nerve center of the city and of a large section of the country. The Leningrad station, both wireless and telegraphic, was of tremendous importance. It helped the Soviet people hold out in a tough battle made worse by the blockade, the famine and the frost. When it was not possible to air the usual program, Radio Leningrad gave out metronome signals. They were a sign that the city was alive. During the blockade, Radio Leningrad broadcasted to the whole country the performance of the Symphony Dimitry Schostakovitch wrote especially for and dedicated to this city. This happened on August 9, 1942. To perform it, the musicians, at that time on the battle front, exchanged their weapons for musical instruments. After the concert they went back to the battle front located some miles away from the city.

In July 1941, the head of the Nazi propaganda, Goebbels, gave the order to silence Radio Moscow. One of the Fascist pilots, shot down over Moscow, had on him a map which showed that among the major bombing objectives was the building of Radio Moscow. On July 22, 1941, a bomb fell into the yard of the radio committee headquarters (at that time in the Pushkin Square in the Capital); but, fortunately, it did not explode. Moscow went on airing programs all through the war, continuously.

Branch offices of the radio committee of Moscow were established in Kuibeshev (Volga), Sverdlorsk (Ural), Komsomolsk-on the -Amur (Soviet Far West). This secured good audibility of the programs in all corners of the country.

On November 7, 1941, a report was broadcasted from Red Square during the traditional military parade that celebrated the anniversary of the October Revolution. German troops, at the time, were at a distance of 25-40 kilometers from the capital. The live broadcast from Moscow's Red Square could be heard all over the country. Shortly after this program, troops left for the front.

The Radio Moscow programs broadcasted within the borders of the Soviet Union were functioning along three main lines:

● radio programs for the front, the army and the fleet, prepared mainly with the help of military and other experts; Among these programs, it is worth mentioning *Letter to the Front, Letter from the Front*;

● radio programs for the partisans; this section started airing systematic programs on June 1, 1942; they included the latest news, information about the partisans' moves, recommendations as to how to carry the fight against the Fascists, letters from family;

● radio programs for the rest of the country, including news bulletins, letters from the front, reports of heroic deeds performed behind the battle lines.

In all sections of the networks, with the exception of the above, the main place was taken by the reports of the Soviet Informational Bureau, various political programs, and concerts on request. Seeing that since 1943 there was an increase in the number of Soviet victories, a special program was initiated. The orders of the supreme commander mentioning these victories were read out. These readings were followed by salvos of the artillery salute. The first such program was aired on August 5, 1943.

Radio Moscow broadcasted a report on the capitulation of General Pauliss the commander in chief of the German troops at the Stalingrad front.

On May 2, 1945, Lazar Magratchev, the Radio Leningrad correspondent, had a live report from Berlin's Unterlindt Street on how the Soviet soldiers hoisted a banner in celebration of the victory over the Reichstadt, and on May 8, 1945, he conducted a report on Germany's unconditional capitulation.

Soviet radio broadcasting for foreign languages during World War II directed its programs to the enemy, Nazi-dominated Germany, and to the allies and the neutral countries.

On Radio Moscow one could hear many voices of the resistance movement, members and leaders of the organizations such as "Free Germany." In his salutation speech sent to Moscow in 1944, the Danish writer Martin Anderson Mekje said, "The reactionary forces are starting to shut up radio broadcasting, to silence the words of freedom, but this will never be possible. All who have ears to hear can tune to the known wavelength, regardless of everything else. In partisan thickets, in far off bunkers, up in the sky and deep under water, everyone can hear the voices of the free radio. Here in the Danish underground, we listen to Radio Moscow and Leningrad. The Nazi jammers roared on the air like wild beasts; but, in spite of everything, we listened to the voice of the truth and found out about the latest victories, about the crushing defeat of the German Fascist troops near Moscow, of the extraordinary endurance of Leningrad, and about the Stalingrad victory. The waves brought news to us of the fight of others and of their news. While fighting for freedom, mankind came to have a common dream. Thanks to the radio, we, all over the globe, gave each other a brotherly handshake."

During World War II the radio organization of the allies, the Soviet Union, the United States and Great Britain, collabrated to broadcast joint programs. Radio Moscow relayed to the Soviet audience programs from England and the U.S.A.; in their turn, English and United States listeners could tune to translated programs from Moscow. On September 21, 1941, two big concerts of Russian and Soviet music were broadcast for the English and U.S. public. In 1943, at the request of the American "Blue Network," Radio Moscow started a special weekly program for U.S. listeners. For that, musicians and huge choirs, such as the Krasnoznamensk Ensemble, the Singing and Dancing Ensemble of the Red Army, gathered in the concert hall at 4 a.m. so that the time would be convenient for a U.S. audience to listen to Soviet and Russian songs.

In the postwar years the development of radio broadcasting was as intent as before. But now it was seconded by a parallel development of television. At the beginning of the '50s it was commonly thought that a brisk growth would mark the end of radio broadcasting. This did not happen, however, and radio did not lose its position.

The postwar years were characterized by a steady, evident and international process of growth in the number of radio stations and of the total volume of broadcasting hours. There was a boom in the development of wireless and telegraphic broadcasting. The telegraphic system (relaying transmissions to inhabited localities) secured a high technical quality of reception, exceptional for a country as big as the U.S.S.R. The reception devices of radio relay centers in cities and smaller urban settlements servicing certain groups of subscribers were of a higher quality than the usual average type of radio receptors.

Already by 1948, the central network resumed its three program transmissions. The daily volume of their broadcasting exceeded 45 hours. Each republic, county, and district had its own programs. The number of telegraphic relay stations increased by 9685 throughout the country from 1945-1950.

The All-Soviet Committee in charge of radio broadcasting and equipment with radio stations, affiliated to the U.S.S.R. Minister's Council, was reorganized in July, 1949. A radio information committee, affiliated to the U.S.S.R. Minister's Council, was created for broadcasting to foreign countries. In 1947, programs for the U.S.S.R. were aired in the 40 languages of its nationalities, and they averaged three to four hours a day. This did not include local broadcasting. More than 100 radio stations were in operation. Their programs were received and relayed by 7000 radio points and 5 million telegraphic transmission points.

On May 16, 1957, the U.S.S.R. Minister's Council created a state committee in charge of radio and television, affiliated to the U.S.S.R. Minister's Council, combining radio and TV broadcasting. Within the committee, several boards were created: the board for inner broadcasting (starting with 1959 there was also a board for broadcasting to foreign countries), one for artistic programs, one for television, one for broadcasting on location, one for the exchange of radio and TV programs with foreign countries, one for technical services, another for financial matters, and so on.

Radio broadcasting continued to grow during this period; there was an increase in the broadcasting hours and progress in the quality of the programs. In 1957, a new radio network was created. The first and second programs started airing a whole variety of shows. Their topics were universal and of common interest. The third program mostly specialized in dramatic, artistic, and musical programs. There were programs aired on ultrashort wavelengths. These programs specialized in broadcasting musical programs.

A landmark in Soviet radio broadcasting was marked inside and outside the country by the Eighth Youth and Student Festival held in Moscow in the summer of 1957. The abundance of events and their variety required a huge effort and great mastery on the part of newspaper and radio workers.

The central radio station started its broadcasting in 1958; later it became the fifth program. It was a program for foreign countries in Russian, directed to Soviet citizens and persons of Russian origin, as well as to foreign citizens who were familiar with or learning Russian. Starting August 1, 1960, a fourth program was started.

In 1964 (June 24), a 24-hour news and music program was created, called *Mayak*, which was broadcast all over the country. It has 48 news updates at every 30 minutes. *Mayak* was the second program of the central network. From a program with a general character it turned into a highly efficient and specialized service. Its creation was conditioned, along with other causes, by the growth of general interest throughout the country in an efficient information service, as a result of the great mobility of the audience, the large span of the territory, the numerous temporary economical zones, and also the competition of television. In that same year, the number of broadcasting hours on the third program increased to five per day.

In 1967, the creation of the multi-program telegraphic broadcasting was started. Three basic multi-national programs, as well as local programs in the republics, began to be relayed through one pair of cables to the radio sets of the subscribers. By 1970, three radio programs were already transmitted through more than 5 million relay centres.

On October 15, 1962, Radio Station *Youth* started its programs directed to young people. As mentioned before, broadcasting for the youth started toward the mid '20s; but the logical results of these programs, the young people's growing interest for broadcasts directed to them in particular, indicated the need for creating the independent station *The Youth*. This station had several sections dealing with the creation of specially-oriented programs for different categories of listeners as to age and profession.

TELEVISION DEVELOPMENT

We will have now to turn back to the mid'30s so as to follow the creation and development of television in the Soviet Union. The first television program was broadcasted from Moscow on November 5, 1934. It was a 25-minute concert in which actor Ivan Moskvin read Tcheckov's short story *The Criminal*, followed by singers and ballet dancers.

Television at that time had a very limited range of broadcasting. The first programs were basically variety concerts and dramatic performances (plays and operas). However, the lack of professional expertise within mechanical television broadcasting did not allow for it to develop into a large-scale mass medium, even though the material stimulus for further work in this field was there.

Television broadcasting of a more expert kind was started in Moscow in March, 1939. Approximately at that same time, it was started in Leningrad as well. As before, the programs were mainly dramatic performances, variety and symphonic concerts and short shows on public and political themes. In November, 1939, Moscow television broadcasted programs dedicated to the jubilee of the First Cavalry which became famous during the civil war.

In 1940, TV programs included more and more news bulletins read by TV announcers. Such programs paralleled the news updates released on radio stations. Besides the news, television programs also included movies and live performances.

The beginning of World War II hampered the further development of television. Its revival on a new technical basis started only after the war.

TV Moscow was the first in Europe to resume regular programs on October 15, 1945. At that time there were 420 TV sets in the houses of Moscovites. On November 4, 1948, TV Moscow broadcasted an experimental program with 625-line image. This has since become a standard.

By the beginning of the '50s, television development was held back by a complexity of technical problems related to the transmission of programs, as well as a lack of receivers and prospective receivers. In 1949, the construction of a comparatively powerful 8-inch diagonal TV receiver was started. Still, in 1949, TV Moscow created the first mobile TV station.

In the early '50s the development of television became a state objective. The building of TV centers and the installation of TV receivers were included in the state plan of national economic development. In 1954, there were 225 thousand TV sets in the country, which meant up to 1 million viewers per day. Special interest was aroused by out-of-studio TV reports.

On November 6, 1954, a program was released about the festive conference held in the Bolschoi Theatre and dedicated to the 37th anniversary of the Great October Socialist Revolution. In the summer of 1955, a program was shot at the All-Russian Construction Exhibition and the All-Russian Agricultural Exhibition. On May 1st 1956, the parade and demonstration of the working people was televised in Red Square. From that day on, on the 1st of May and 7th of November each year, programs have been broadcasted from the Red Square. Now they are transmitted through the "intertelevision" system, and often enough they are broadcasted by other television systems for hundreds of thousands of people throughout the world.

An important event in the development of Soviet television was the broadcasting of the VI World Youth and Student Festival, held in Moscow in the summer of 1957. The preparation for it and the festival itself came to be a test for the school and a stimulus for the development of television and telejournalism. Each of the 15 days of the festival was advantageously set off.

The growth of the popularity of television, the increase in its audience, the attempt to satisfy fully the various categories of viewers resulted in the emergence of new forms of braodcasting, many of which have since been established as permanent. Such were the popular telejournals as *Art, Knowledge*, and in later years *Movie Panorama, Obvious Impossibilities*, etc.

In 1956, the central TV studio created a section for *The Latest News*, which started out by overlapping with the news issued by radio stations. But, gradually, topics specific to television were introduced into the programs. At first they were televised news. Broadcasting the VI World Youth and Student Festival greatly influenced the development of the informational function of television. The programs of *The Latest News* started being aired daily. Everyday they were more markedly different from the radio news bulletins, *The Latest News* becoming more and more of a newsreel.

In time, it became obvious that filmed news alone could not contain all the current events at home and abroad. A determined percentage of reading by announcers was necessary in order to preserve the integrity of and universality of *The Latest News* on television. Starting in 1960, *The Latest News* program changed its name to *TV Newsreel*.

In the '50s, an important section of television was developed, that of educational programs. The educational competition *Motor Car*, transmitted in 1955, was the beginning of these programs. It was followed by the series, *The Origins of Life on Earth*. This program included not only lessons, material of factual character, but also roundtable discussions.

The year, 1957, was important for the emergence of a highly popular quiz show called at first *Funny Questions Evening*. In a couple of years it was turned into a program called *The Fun and Wit Club*, which in several years was broken into several new shows of the quiz type.

A necessary and inherent part of television programs were variety shows, first live performances and later on taped performances. This was the foundation of genuine television broadcasting, which has since grown in three main directions, corresponding to the uses of television and, lately, of magnetic recording: artistic, informational and scientific.

In the first stage of television development, one could sense the influence of those mass media which preceeded television: film, drama, radio. These informational and artistic modes had their own laws as well as problems. Their synthesis started in the '50s, and was a process of crystallization of television as a mass medium and an artistic mode.

The next years were years of brisk development for television; the number of TV studios grew in the Soviet Union; there was a steep increase in the number of TV receivers and viewers. In 1955, there were TV stations in the country, and in 1960 there were 84 studios. In 1955, the volume of the broadcasting was 15.3 hours daily and in 1960 it was 276.5 hours daily. In 1955, the population owned 823,000 TV sets and in 1963, 10,400,000. In 1955, there were 402 workers on TV studio payrolls, and in 1960 the number rose to 8295. These people had to master everything that was new not only for themselves but for the sake of artistry in their whole profession; this could, of course, but result in a quest for highly divergent, successful as well as less successful forms of TV broadcasting.

Television wholly involved itself in the problems facing the people and the state. During the process of development of television it became clear that the mere existence of the TV screen does not make television as such. It was proved that television can easily become a radio receiver, if the screen does not provide visual, graphic information. It became obvious that television broadcasting needed specific forms and structures corresponding to both its function and principal modalities.

Parallel to the qualitative growth there was a qualitative improvement of television. The structure of daytime programs became more and more precise, shows that were addressed to various categories of viewers. During the evening, television monopolized the audience. This is largely due to the fact that at 9 p.m. there is a full *Newsreel*. Since 1960, *Newsreel* has been aired daily. On January 1, 1968, another news program was initiated, resulting in structure to a full-weight daily chronicle called *Time*. This show provides an original viewpoint and an individualized report within the structure of the daily shows. During 24 hours the first central television station aired two issues of the *Time* show, at 8 a.m. and 9 p.m. Besides that, there are news bulletins in the form of roundtable talks led by the commentator of *Today in the World*, as well as special issues.

Toward the mid '60s the social effects of such mass media as radio and television became apparent in a more definite form than before. With the emergence of television in the '60s there was a sensible change in the structure of the time expenditure of the consumer living in a developed socialist society. The data provided by sociological studies carried on in the Moscow University, Leningrad and other cities, yielded highly interesting results about the division of time between various mass media. The most obvious direction, as indicated by these figures, proved to be an absolute growth in the time used up by mass media. This can be attributed to both the greater number of TV sets in the possession of the people and the increase in the population's free time. Some researchers find that the amounts of time spent in reading newspapers and listening to the radio have a definite level of consistency.

Another consistent effect of the social influence of radio and television broadcasting is that modern man is taking in information from several sources simultaneously at varying levels of perception. Radio and television stand apart as far as the perception of programs is concerned within the framework of one day or 24 hours of activities. Radio, and even TV at times, is often received as a background medium.

The appearance of television and its diffusion into the masses did not have a negative effect on the development and popularity of radio. On the contrary, radio broadcasting was further developed and perfected during the rise of television, which must be attributed not to the weakness of television but to the specific perception radio programs elicit. According to statistical data, in 1965, 70.9%, and in 1967, 76.1% of the population listened to the radio daily.

Within a 24-hour cycle, radio listening has several peak hours. The morning one is during preparation and departure for work, from 6 to 8:30 a.m. The second one is at lunch hour, 12 noon; and, finally, the third one is after 6 p.m. (However, at this time television regularly takes over the audience, capturing their attention from 7-11 p.m.)

Today, the actual radio audience represents up to 50% of the population. The higher morning peak hour counts up to 60% of the workers and clerks. And another observation: watching TV programs primarily falls into the free time of the population, while listening to the radio programs (sometimes very short) occurs often enough during work hours, as well. Connected with the fact that in the Soviet Union certain sections of the industrial and administration staff work in two or three shifts, many of the more interesting TV shows usually aired in the evening are repeated in the morning for workers in the second shift.

Man's total dependence on information is especially manifest in the field of utilitarian information, providing primarily for the everyday routines, as well as contributing to the political, social and entertaining aspects of life.

According to a series of Soviet researchers, one can divide the attitude of persons toward sources of information into three categories: spiritually- and privately-oriented, professionally- and functionally-centered, and practically-oriented. In any case a person's involvement with mass media channels can be qualified only when knowing the character of his or her work and place in the social structure.

It turned out that an essential datum in the activity of mass media was the fact that the content matter of radio and TV programs was not aimed at the alienation of people from their social environment and did not result in their social and psychological isolation. The content matter of mass media programs must strengthen the links between the listener and the life of the community, must call on him to play a more active part in the activities around him.

Soviet Researcher R. A. Boretsky noticed a set of peculiarities of television in the Soviet Union and other Socialist countries. After studying the content, genre and thematic characteristics of programs in four socialist countries, he reached the conclusion that socialist television is directed in the first place to the instruction and education of the large masses, differing in this basic trait from television in capitalistic regimes. In the Soviet Union, educational programs make up around one-third of the total volume of television broadcasting.

Television in bourgeois countries, more than that of socialist countries, has entertaining objectives. Such programs commonly occupy up to one third of the total bulk of broadcasting, or sometimes even more times the amount of time devoted to educational programs.

In recent years, Soviet radio and television have been intensely developed, and their audiences have continually increased. In 1973, 83% of the U.S.S.R. population, including that in the far north, Siberia and far west, had the opportunity to tune to Moscow programs. There are now more than 70 million TV sets throughout the country. The number of color television sets is on the increase as well, so that, if compared with 1976, in 1978 the number of color television sets has doubled.

In addition to the central station, in national and autonomous and republics, in counties and districts, there are 122 television centers, each with good technical facilities. Besides that, there are 140 powerful broadcasting television stations in the country and around 2000 relaying centers.

The huge territory and the widespread population of the U.S.S.R. necessitated the development of satellite television. The Soviet Union was the first state in the world to create a system of satellite television, capable of receiving television broadcasts from practically all over Soviet territory. The difficulty of creating the national cable system, because of the large territory to be covered, made it imperative that satellite television should be created.

The foundation of Soviet satellite television was constituted by the satellites "Molnya" and "Ekran," and the "Orbit" ground station, and the network of receiving centers built for common use. At the beginning of 1979, 82 "Orbit" type stations were functioning in the Soviet Union. Through this system, and using the Soviet "Molnya" satellite, the U.S.S.R. facilitates exchanges of television shows with socialist countries such as Bulgaria, Hungary, East Germany, Cuba, Mongolia, Poland, Czechoslovakia. The exchange of programs with more powerful television stations of other countries is made through the InterTelevision and EuroTelevision systems.

The state plan for economical and social development for 1976-1980 anticipates broadening the areas of high-quality reception of television and radio programs for all parts of the country, the further development of satellite connections, as well as the development of color television and stereophonic radio broadcasting. Since 1978, all of the programs of the central station are broadcasted for color TV. As a matter of fact, 33 cities broadcast their programs in color, the capitals of 15 Soviet Republics included, as well as Leningrad, Sverdlovsk, Novosybyrsk, Tchelyabinsk, Bolgograd, Saratov, Vladivostok, Simferopol, Sotch, Kuybishev, Omsk, Dnepopetrovsk, etc.

In Moscow the Olympic Teleradio complex will secure the broadcasting of the full load of news on the XXII Olympic Games to all interested countries in the world. The equipment allows for broadcasting 20 color television and 100 radio programs for foreign countries. The preliminary data show that the Moscow broadcasts for the 1980 Olympic Games will be watched by 2 billion people, half a billion more than during the 1976 Olympic Games in Montreal.

Stereophonic radio broadcasting is being developed, and it is already used by 30 cities. Another 10 radio stations recently began stereophonic broadcasts.

CENTRAL TELEVISION NETWORK

Central television carries its programs on seven channels. Three of them have repeats of the first channel.

The First Channel: The main one, its programming covers the whole country and has an informational, socio-political, scientific and artistic character. The average daily broadcasting volume is 13.5 hours per day. A great part of it is dedicated to cultural programs, movies, and concerts. Literary and musical shows occupy up to 40% of the total hours. A great part, up to 60% of the broadcasting hours, is devoted to informational, socio-political, scientific and popular shows. They include shows directed to a specific section of the audience: teenagers and children, for example. One can include here also the numerous sports programs.

The Second Channel: Directed to the Moscow public, though it can be seen in the neighboring areas as well. Its programming is informational, journalistic and artistic. It operates 4.5 hours daily.

The Third Channel: This is the scientific, popular subject and educational channel. It includes programs for students and teaching staff of secondary schools, as well as those in higher education. Some shows are directed to state specialists, teachers, and doctors. This channel also has scientific and popular subject series for a large audience and feature movies with subtitles for people with hearing impediments. At present the area served by the third channel has been enlarged. Apart from the Russian Federation, it can be watched in the towns of Ukrain, Bel Litv, Latry, and other areas. More than 23% of the Soviet population can watch this channel. Its average daily broadcasting time is 6.1 hours.

The Fourth Channel: Programming on this channel has a socio-political, artistic and athletic character. The daily broadcasting average time is seven hours. Thanks to radio relays and cable lines its programs can be watched in seven autonomous republics, in the Krasnogarsky county, in 34 sections of the Russian Federation, in Ukraine, Bielorussia, Moldavia, Latviskia, and Litovskia. Through the satellite system it can be watched in the cities and villages of the middle east, the Kau Kose, the Black Sea coast, Ural, Siberia and the far west. The audience of this socio-political and artistical channel is constituted by more than 62 million people. At the end of the 10th 5-year State plan (in 1980), the areas where the 4th channel will be received will be even larger.

As there are two time zones in our country from the Beringov Straits to the Baltic, the first programs of the All-Russian channel are transmitted through satellite and through the "Orbit" system, timed so as to be received by the people in Siberia, middle east, far north and far west. Each broadcasting area covers several time zones. So, "Orbit I" is directed to the viewers in Tchukotka, Kamtchatken, Sachalin and Magadansky. The daily average broadcasting hours is 13.5; the difference in time in these districts is from eight to 10 hours.

"Orbit 2" is directed to the districts of eastern Siberia and Primoria (far east). The daily broadcasting hours average 13.1. The difference in time between Moscow and these districts ranges from five to seven hours.

"Orbit 3" ("Vostox") is directed to the inhabitants of western Siberia, middle east and Kazachstan, besides the Ural and far north. In order to transmit repeats fo the first channel program to these districts, a satellite and ground relay system is used. The daily broadcasting volume averages: for "Orbit 3," 13.1 hours, and for "Vostok," 13.5. These systems serve the areas which have a time difference as to Moscow ranging from two to four hours.

NATIONAL RADIO NETWORK

The National Moscow Radio Network offers daily broadcasts on eight program channels for a daily average of 158.6 hours.

The First Program is the main, all-Soviet, socio-political and artistic channel. The daily average volume of broadcasting consists of 20 hours, 50% of which are taken by news, socio-political programs, shows for children and young people. An equal amount of time is taken up by dramatic, literary, musical and athletic broadcasts. Due to the difference in time, the All-Soviet radio airs variants of the first program for the western Siberia, the republic of Middle Asia and Kozachstan, for eastern Siberia and for the far east. Each of them has 20 broadcasting hours daily.

The Second Program, "Mayak," mentioned before, is a round-the-clock, informational and musical program. It is aired simultaneously for all of the regions in the country. It contains news bulletins aired every 30 minutes, and in the morning hours even more often, announcing the latest events in national and international affairs. The bulletins last four to five minutes. "Mayak" broadcasts the best works fo Russian, Soviet and World symphonic, popular and live music.

The Third Program of All-Soviet radio is a literary and musical program. The average daily broadcasting hours amount to 16. It airs plays for the radio, literary shows, classical musical works, symphonic and operatic, programs on the life and creations of Soviet and international writers, musicians and artists.

The Fourth Program is musically-oriented. It is aired on medium and ultrashort wavelengths for nine hours. International classical works are aired in stereophonic variants. This program is directed to lovers of symphonic music and is notable for the high quality of the sound.

The Fifth Program operates around-the-clock, providing musical and informational programming. It has socio-political as well as artistic shows and is directed to Soviet citizens abroad, as well as to sailors, fishermen and polar explorers.

Radio broadcasting covers practically all the territory of the Soviet Union. On January 1, 1979, the U.S.S.R. listeners had at their disposal around 68 million radio receivers and around 70 million radio relaying stations that were functioning. In the use of relaying stations a tax is percepted, which amounts to 50 kopecks per month (80 cents). Radio and television sets subscribed are not taxable.

Besides the All-Soviet radio broadcasting network, there are local radio stations in each Soviet and autonomous republic, county, district, and in each city and regional capital in the country. Their programs are carried in 69 languages of the peoples of the Soviet Union.

Central radio broadcasting is directed to foreign countries (Radio Moscow) as well as the radio station of the "World and Progress" organization members who air programs in 69 languages to 150 countries in the world.

INTERNATIONAL BROADCASTING

A significant turn in Soviet radio broadcasting to foreign countries was marked by the creation of a World Service in English. Its programs were first aired on October 3, 1978. It operates 19 hours daily and is directed practically to all countries and continents. The principal task of the Radio Moscow World Service is to air around-the-clock news on world and Soviet affairs. A full news bulletin is aired every hour. Every 30 minutes there is a Soviet news update. The programs include reports and comments on international issues and of the life in the U.S.S.R. Permanent programs of the World Service are *Radio Update on Current Affairs, News and Opinions, People and News*, etc. They are repeated every four hours and at convenient listening hours. The second 30 minutes of each hour is taken up by entertaining and educational programs. Listeners can familiarize themselves with Soviet culture, travel through the U.S.S.R., listen to symphonic, contemporary, variety and popular music.

The radio programs of the World Service also include interviews with guests of the Soviet Union and with Soviet people. Foreign listeners get sufficient information about the life in the U.S.S.R., as well as its international policy. Only three months after the launching of this program, letters were received from 47 countries (USA, Greece, Spain, England, Philippines, West Germany, Iran, Japan, Poland, etc). The great majority of listeners approved of these programs, underlining the fact that now at practically any time of the day they can get efficient and many-sided information and find out what the Soviet Union point of view is on international affairs.

Due to its traditional international policy of friendship and collaboration, and its implementing of the provisions of the European Conference,

Final Document, Soviet radio and television maintains connections with organizations in other countries and exchanges information with them. In 1978, within the exchange of programs and movies, there were joint shootings, festivals, symposiums with 136 television and 117 radio stations in 120 countries.

In 1978 in particular, central television showed 13 movies and 400 other programs of U.S. make. Together with the C.S.A. Company, 36 programs were made jointly, and there were more United States delegation visits to the U.S.S.R. Soviet television and radio maintain a very sustained collaboration, especially with socialist countries and organizations.

THE OUTLOOK FOR FUTURE GROWTH

The intensive development of radio and television broadcasting in the last several decades, the substantial changes that this development brought about in the life of the community and the significant role played, has induced researchers to raise realistic and practical enough questions about the future, about the place of mass media in the coming years. The prognostication of the way the mass media will develop is complicated by the absence of some basic and relevant data on the direction of their evolution. The prediction is further impeded by the existence of certain illusions, generated by an imperfect knowledge of the potentialities of mass media.

In raising this question one can distinguish between two aspects. First, there is an international one, the activity of interstate mass media, in particular radio and television from and for foreign countries, including radio and television via satellite. The second aspect is the inner, national activity of mass media within each country.

As for the international aspect, one can say that the intensive development of television and radio broadcasting will be carried on. One can suggest, with a certain amount of trust, that the tendency toward a brisker development of radio independent from the emergence of television will be continued. The perspective of radio and television broadcasting from one country to another is conditioned greatly by technical progress. One can presume, however, that a more important place than before will be granted to the technical problems of mass media (radio and TV). The emergence of television in the form of live broadcasting on the international arena will not diminish but possibly alter the important part played by radio broadcasting in the world.

Obviously two factors in the development of television, but partly radio as well, will compete and interact in the future. Cable television is convenient for states whose territory is comparatively small, where the expenses of setting up the cables is taken care of by the density of population in the serviced regions.

The diffusion of cable television obviously offers the possibility of a two-way connection between viewers and the source of information, not only the TV station, but also information from book and film libraries. Such information can be given out in convenient forms such as records, cassettes, etc. But, such a television system isolates the audience from direct contact with satellite television; it does not allow them to choose the programs aired in the world.

Satellite television in particular, direct television broadcasting, which will actually become a reality in the mid '80s, would solve such questions as providing the audience a comparatively inexpensive means of access to television programs. But this raises new problems connected with the fact that it opens the way to an unimpeded penetration of TV programs into the territory of other countries, which would prevent the development of domestic television.

To solve these problems there must be international cooperation between the states, a definite set of norms and rules must be established as to the utilization of mass media in the interest of the whole world and of the improvement of friendly, neighborly relations. It is to solve these questions that in August, 1972, within the UNO Project, a convention was held in the Soviet Union entitled "On the Principles Governing the Use of Different States of the Artificial Satellites of the Earth for Direct Television Broadcasting." At this convention the Soviet Union suggested a set of regulations under which each state must carry on its broadcasts with a view to the interests of the world as a whole. Where a violation of the accepted norms occurs, the wronged states would be granted the right to use available measures of counteraction.

A similar convention could have regulated the use of television, subordinate it to international interests, to the exchange of cultural and national values, and the strengthening of friendship between countries. These problems are tackled in the various sections of the Final Act of the European Conference on Security and Collaboration held in Helsinki in 1975.

On the national level, the development of radio and television broadcasting in the U.S.S.R. obviously will take directions evident in the layout of the state plans for the development of the Socialist economy for the near future. At the end of 1980, the percentage of the U.S.S.R. population with access to television will increase to 85%. The system in existence today ("Orbit") will allow for two-way television broadcasting; i.e., there will be an increase in the number of TV stations broadcasting efficient programs from Moscow within the central network and which are further transmitted from the capital throughout the country. The connective system with the satellite Intelsat will be utilized, which will better the interstate exchange of TV programs, and various systems of video recording on cassettes and records will be intensely developed.

At the end of 1990, the percentage of the U.S.S.R. population with access to television will be 95. Three programs of the central station will become All-Soviet programs (now only 1 is). In the republics, they will be supplemented by a 4th, republican, and in big cities by a 5th, metropolitan, program.

The first and second channels of the central TV station will be transmitted (starting as early as 1980) thru five and not three transmitting channels, which will be a definite improvement in the servicing of audiences in different towns. The third All-Soviet channel will be repeated at a time convenient to the pace and habits in the life of people in all of the time zones.

There will be wide distribution and an increase in production of the kind of gear used for recording and reproduction on the screens of standard TV sets. Recordings of the 3rd All-Soviet educational programs will be widely diffused in the population.

The combined use of satellite systems and cable television gear will facilitate the accomplishment of a multi-purpose informational service for the population, including separate demographic groups. This will make it possible to carry on an exchange of programs between local republics and the central TV station. In this manner, the perspective growth of television in the U.S.S.R. is secured with regard to the following three statistical objectives:

1. The first objective will provide a maximum section of the U.S.S.R. population with TV and radio programs. This objective is already reached as far as radio is concerned. Nowadays, TV programs from Moscow are transmitted through a cable system and radio relaying channels to all the cities and regions of the U.S.S.R. with a significant number of inhabitants. Until 1975, the three-fourths of the U.S.S.R. population with access to television occupied only one-tenth of Soviet Union territory. The remaining fourth without access to television was distributed over nine-tenths of the country's territory.

The transmission channels based on the use of satellites include a number of variables. The "Orbit" system was the first used to service a large territory. It was able to function only with complex and expensive relaying equipment, located in big and middle-size towns in Siberia, far Vostok, and far north. The "Ekrau" system, made afterwards, would make use of collective antennas, comparatively inexpensive to build, located in small urban and rural areas.

2. The second objective is that of taking into consideration the demands of the audience and, taking into account the existence of many time zones, to supply TV programs precisely computed for the local time and life pattern of the population. This implies the dislocation in time of different regions. This entails the necessity of taking into account the programs of local stations as well as the important events that are taking place during the broadcasting. For such broadcasting, ground and satellite systems are used.

Ground channels serve the central region of the Soviet Union, as far as the Urals. From there on, the programs are transmitted without any time lapse.

The programs for the regions east of the Urals, western Siberia as well as for the territory of the middle east republics: Turkmeny, Uzbekistan, Kazakistan, Tadzikestan, Kirghizy, are transmitted also through ground channels, but with no time lapse. This repeat program is called Vostok/East.

Finally, the regions lying further to the east are serviced exclusively by satellite channels with no time lapse. State plans for the economic development of the country evisage the increase in the number of programs transmitted with no time lapse to five broadcasting areas. The difference between the eastern and northern areas will be no greater than two hours.

3. The task of completely overcoming the difference between urban and rural areas, between the economic and cultural levels of development of central and faraway regions, constitutes the third objective—that of expediently securing a multichannel TV program for all these regions— the far-off ones included. For the fulfillment of this objective it is imperative, by 1990, to secure access to at least three all-Soviet TV programs in all the regions of the U.S.S.R.: the first channel (information-al, socio-political and artistic) with a daily total of 16 broadcasting hours; the second (socio-political with a wide range of programs) with a daily total of 16 broadcasting hours; the third (scientific and educational) with a broadcasting volume of 12 hours daily. The average daily bulk of republican programs in the Soviet republics is supposed to rise to 13 hours.

These are, in broad terms, the perspectives of the development of radio and television in the U.S.S.R. Besides the technical considerations, there is a more important social aspect. The further development of mass media radio and television, on a national as well as international plane, must lead not only to an even wider scope of man's knowledge and his control over the natural forces, but also to such a utilization of these most powerful means of informing the public and forming public opinion so as to implement exclusively the interests of the whole world and progress on earth. It is under these signs that Soviet radio and television carry on their activities.

Chapter 4
Broadcasting in Japan

Japan Broadcasting Corporation
(NHK)

Radio broadcasting in Japan was begun in 1925 by the Tokyo Broadcasting Station, a corporate juridical body which later became Nippon Hoso Kyokai (NHK—Japan Broadcasting Corporation). The radio network was gradually expanded to cover the entire nation, and broadcasting became part of everyday life in Japan.

In 1950, when Japan was working toward democratization after World War II, the new broadcast law was enacted. This is an important law that cannot be ignored when discussing Japan's broadcasting and its system. It was under this law that NHK made a fresh start as a special juridical body for broadbased public broadcasting on the basis of income from receiver's fees. This law also recognized the establishment of private or commercial broadcasting enterprises operating on the basis of income from advertising. Since that time, Japan's broadcasting has been functioning under this dual system, with NHK and commercial broadcasters operating in parallel.

In 1953, NHK began television broadcasting, soon followed by the private companies. Thus the number of broadcasting stations rapidly increased, and television sets became common in Japanese homes as the nation achieved rapid economic growth. With the advancement to color television, most Japanese became regular television viewers.

Broadcasting thus came to have a powerful influence on the public, and thus it was recognized as an important means of raising the cultural level of the Japanese people.

For example, in the Tokyo area, people can tune in not only to the NHK General Television Service (GTV) and its Educational Television Service (ETV), but also to five commercial television stations. In radio broadcasting, they can tune in to three medium-wave stations and one VHF-FM commercial station, in addition to NHK's Radio 1 and 2 and VHF-FM broadcasting networks. For Japan as a whole, NHK's GTV and ETV and commercial broadcasters each operate more than 2700 television stations, for a total of about 9000, in order to cover the entire mountainous nation.

THE INFLUENCE OF TELEVISION

According to a 1975 survey by NHK, 93% (95% on Sunday) of the Japanese people watch television at least once a day, and the average viewing time per person is 3 hours 19 minutes a day (4 hours 11 minutes on Sunday). See Table 4-1.

In the same survey, those who said that television was useful to their home management and social life totalled as high as 96%. The great majority said that television has a significant cultural and educational influence, allowing greater understanding of overseas conditions, spreading new ways of thinking, and conserving the traditional culture of Japan.

When the respondents were asked which media they would prefer for obtaining news, about two thirds of the total picked television. This is more than twice the number of those who indicated newspapers as their preferred source. The survey results follow:

Television..64%
Newspapers...25%
Radio...10%
Conversation 1%
Listen to wire broadcasting.. 0%
Weekly magazines 0%

These survey results clearly show how much the Japanese people depend on television as an information source.

NHK: JAPAN'S PUBLIC BROADCASTING ORGANIZATION

NHK, the only public broadcasting organization in Japan, operates networks that cover the entire nation. Thus, it exerts a far-reaching influence on the Japanese people, and care must be taken that NHK will not be used for political purposes by the government or for business purposes by private enterprises. Its programs must accord with the opinions and needs of the general public. As will be discussed later, NHK has a number of institutional guarantees, such as freedom of speech and stable financial sources, to achieve the above purpose.

NHK, as a self-governing public broadcasting organization, is protected by law so as to be independent of all political authority. Japan fell victim to excess nationalism in the past, and this plunged the nation into a disastrous war. But then, embracing democratic principles after the war, the Japanese people were determined not to repeat the bitter experience of restrictions on basic freedoms of speech and the press.

NHK was thus guaranteed, under the 1950 Broadcast Law, freedom of expression in broadcasting, and was directed to "conduct broadcasting for the public welfare in such a manner that its programs may be received all over Japan." The Broadcast Law outlines the following as the fundamental tasks of NHK.

● To broadcast well-balanced, high-quality programs in the fields of news reporting, education, culture and entertainment in order to meet the various needs of the people;

Table 4-1. Average Daily TV Viewing Time of the Japanese Audience.

Average Televiewing Time	Weekday	Saturday	Sunday
1965	2 hrs. 52 min.	3 hrs. 1 min.	3 hrs. 41 min.
1970	3 hrs. 5 min.	3 hrs. 7 min.	3 hrs. 46 min.
1975	3 hrs. 19 min.	3 hrs. 44 min.	4 hrs. 11 min.

● To undertake construction of broadcasting stations even in remote mountainous areas and isolated islands to bring broadcasting to every corner of the country;

● To conduct the research and investigation necessary for the progress and development of broadcasting and to make public the results thereof.

● To foster a correct understanding of Japan by introducing this country's culture, industries and other aspects through the overseas broadcasting service; also, to provide international cooperation, such as program exchange and technical aid, for overseas broadcasting organizations.

Management

In order that NHK can maintain autonomy of management in close relation to the general public without being influenced by special interests, the Broadcast Law provides the following detailed provisions for NHK's management, programs, personnel affairs and finances:

Board of Governors and Executive Organ. NHK's board of governors determines basic policies on such matters as revenue and expenditure, the project schedule and program compilation. The governors are selected from various fields and districts, and their appointments are made by the prime minister with the consent of both houses of the National Diet.

The highest responsible person of NHK, the president, is appointed by the board of governors. The president in turn appoints the vice-president and directors with the consent of the board of governors. The president officially represents NHK and presides over its management.

Deliberations at the National Diet. NHK's budget and project schedule for each fiscal year are reviewed and debated at the Diet and must obtain parliamentary approval. In addition, a business report and account settlement for each fiscal year must be submitted to the Diet.

These procedures are aimed at minimizing direct supervision and participation of the government bureaucracy in NHK's affairs, allowing NHK to operate in accord with the wishes of the public, through a review of NHK's business operations by the people's elected representatives at the Diet.

Freedom of Program Compilation. In order to assure freedom of program compilation, the Broadcast Law provides that "broadcast programs shall never be interfered with or regulated by any person, excepting the case where he does so upon the powers provided for by law."

Receiver's Fee System. Enormous funds are needed to provide a rich variety of programs. Yet NHK receives no funds from the government; its entire expenses are met by equitable receivers' fees. Thus, the public itself can be said to support NHK's operations.

NHK collects these receivers' fees on its own. It is this receivers' fee system that enables NHK to escape government control and to present balanced educational, cultural, news and entertainment programs meeting the diverse needs of the people without regard to program ratings.

The characteristics of NHK described above aim at using broadcasting most effectively for the benefit of the people and at keeping it free from government control and economic interests. This system—the public system—is believed to be the best for accomplishing the above purposes.

Business Organization

Day-to-day business activities of NHK are carried out by four major departments; (1) general broadcasting administration; (2) headquarters for technical administration and construction; (3) general administration for audience service; and (4) staff administration. The planning and drafting of management policies are carried out by the general staff section which consists of eight bureaus. There are also various other divisions with specialized duties, such as the finance department and general affairs department.

With its central headquarters housed in Tokyo's Broadcasting Center, NHK also operates seven regional headquarters and 62 local stations, along with a large number of rebroadcasting and translator stations, all linked up to form a nationwide broadcasting network. Research and development activities are undertaken at NHK's own institutes and laboratories. NHK also maintains overseas correspondent bureaus in 21 major cities throughout the world.

Broadcasting Center

The NHK broadcasting center in Shibuya, Tokyo, is the vital core of NHK's nationwide and overseas activities.

Construction of the broadcasting center was started in 1963 in order to meet the broadcasting requirements of the 1964 Olympic Games in Tokyo. The center was expanded thereafter and finally completed in 1973, when all NHK's broadcasting facilities and offices were concentrated there.

The broadcasting center (Fig. 4-1), built on a site of 82,000 square meters, is a 23-story building with two large 8-story wings. With a total floor space of 176,000 square meters, it houses all the program production, transmission and administrative operations of NHK in Tokyo.

Program production and transmission facilities are concentrated in the two wings, while the administrative offices are located mainly on the highest floors. Twenty-two television studios and 23 radio studios with related program production facilities and program transmission equipment for each network are so arranged as to allow the most effective program production and transmission. Operations are facilitated by NHK-TOPICS (Total On-line Program and Information Control System), which is a fully computerized system for aiding program production and automatic program transmission. NHK now produces some 1700 programs a week, including those for both television and radio. The production of these programs, of course, requires diverse facilities and close cooperation of many staff groups, with efforts made to achieve originality in each and every program.

To ensure smooth and effective production work, rapid and accurate information processing and exchange are essential. In TOPICS, such items of information as time schedules, requirements for production facilities, and the necessary number of personnel are processed and sorted and then sent directly to the related departments by the computerized system on a real-time basis.

TOPICS consists of two subsystems: (1) SMART (Schedule Management and Allocating Resource Technique) which arranges the necessary schedule for each program production and other relevant matters, informs the groups of the staff concerned, and automatically allocates facilities and equipment for each program; (2) ABCS Automatic Broadcasting Control System) which carries out the automatic switching and controlling of production and transmission equipment. Since its introduction in 1968, TOPICS has played a pivotal role in NHK's broadcasting activities.

NHK Hall, located next to the Broadcasting Center, is a multipurpose auditorium with a seating capacity of 4000. It is designed to present operas, concerts and other public performances. The architectural acoustics of the Hall were designed by NHK Technical Research Laboratories. The Hall is fully equipped with advanced theatrical facilities, including a computer-controlled stage lighting system, as well as with complete broadcasting facilities for television and radio. Several live or recorded audience-participation programs are produced here every week, and many public performances by artists from abroad, as well as monthly concerts by the NHK Symphony Orchestra, are also given at the Hall.

Programs

NHK has established the Standards of Domestic Broadcast Programs under the advice of the central broadcast program council, which is composed of persons of learning and experience appointed by the president with the consent of the broad of governors. These standards, with the following principal objectives, serve as day-to-day guides in program production: (1) To contribute to achieving the ideal of world peace and to the happiness of mankind; (2) To respect fundamental human rights and effectively promote the spirit of democracy; (3) To promote the elevation of human character, higher cultural attainments, higher

Fig. 4-1. View of the NHK broadcasting center (courtesy NHK).

sentiments and ethics, and rational thinking; (4) To preserve Japan's traditional culture and to foster and disseminate new cultural achievement; and (5) To maintain the authority and dignity of public broadcasting and to meet the expectations and needs of the public.

NHK also adopts basic plans for each year's domestic broadcasting programs. These plans serve as a basic guide for programming and the planning and production of individual programs.

In programming, the results of four scientific surveys (namely the people's time budget survey, the program opinion survey, the program rating survey and the public opinion survey, all undertaken by the NHK Public Opinion Research Institute) are used as basic data for compiling programs which will better meet the requirements of as many citizens as possible.

NHK's domestic broadcasts are presented over five networks, including two television, two medium-wave radio and one VHF-FM to offer the public a wide range of choice.

General TV and Radio 1 Networks. In these services, news, cultural and entertainment programs intended for the general audience are the leading items. While presenting such programs under well-balanced scheduling, these two networks also take care of local services.

Educational TV and Radio 2 Networks. These two services deal mainly with educational programs, especially for school broadcasting. All programs on GTV and ETV are presented in color. Program compilation for the two television networks is flexible. For instance, a cooking lesson presented on GTV in the morning for stay-at-home housewives is repeated on ETV in the evening for working housewives.

FM Network. Full advantage is taken of the high sound quality and limited service area to present many stereophonic music programs, and local programs. Each program undergoes review by NHK's program monitoring staff, and audience reactions are collected from panels of listeners and viewers recruited by public invitation.

News Programs. NHK's news programs provide the nationwide audience with an accurate and objective information service. These news programs not only follow current events stage by stage buy also analyze background information and the main issues involved. Some 1000 reporters and news cameramen based at NHK's broadcasting stations throughout Japan, 25 Tokyo-based news commentators, and 41 foreign correspondents deployed in 21 major cities throughout the world devote themselves to providing rapid and comprehensive service for both radio and television.

Regularly scheduled news reports are presented 11 times on GTV between 6 a.m. and 11 p.m. each weekday. The major news reports are presented at 7 a.m., 12 noon and 7 p.m. On the Radio 1 Network, news is presented every hour on the hour. News is also presented on the VHF-FM service.

In addition to these regular news transmissions, GTV presents a morning news and features show *Studio 102*, and an evening "news magazine" program, *News Center 9 p.m.* These programs offer vivid and flexible coverage of current topics.

NHK's news programs naturally serve as a valuable source of information. Table 4-2 indicates the news sources the Japanese people use for obtaining information.

Table 4-2. Popularity of Various News Sources With Japanese Audience.

News Sources	Regular news	News on major events	Information on typhoons, traffic strike etc.
NHK TV	81.6%	85.8%	83.7%
Commercial TV	47.4	42.8	41.8
Newspaper with nation-wide circulation	46.5	41.5	29.2
Local newspaper	29.3	24.0	17.5
NHK radio	13.0	9.1	22.1
Commercial radio	12.4	7.8	14.6
Others	1.3	1.2	1.4

Side by side with these news reports, film documentaries and discussion and interview programs deal with questions of nationwide interest, such as environmental pollution, traffic safety, energy resources, food supply and economic problems.

In the case of an earthquake or typhoon, NHK is obliged to broadcast accurate information to help insure the safety of the general public. Such activities by NHK in emergencies are designed under the Basic Law for Measures against Disasters.

It is another duty of NHK to relay parliamentary debates over its television and radio networks in order to inform the public and thus help build a sound basis for Japan's democracy.

At times of parliamentary elections, each candidate broadcasts his political views from NHK's stations in each electoral district. Election returns are speedily and accurately broadcast by making full use of a computer system.

Local Broadcasting. The local service on GTV presents mainly prefectural news, information and topics of local interest. Local services on the Radio 1 and the FM service present similar information along with music. Traffic information and weather forecasts are also important items valued by the listeners.

Sports Programs. Sports programs are presented for an average of 570 hours a year on television and 610 hours on radio. The most popular sports such as baseball and sumo wrestling are given priority, with special attention also paid to the promotion of amateur sports.

Educational Programs. Programs for school education and adult education broadcast over nationwide networks are an effective means of equalizing educational opportunities. For this purpose NHK has been expanding its social education programs by presenting lectures and courses on a wide range of subjects. In 1931, NHK inaugurated its Radio 2 Network for the purpose of presenting chiefly educational and cultural programs. In 1959, NHK also established its ETV and began broadcasting organized series of educational programs.

NHK now presents school broadcast programs intended for kindergarten, primary school, junior high school and senior high school use on its ETV and Radio 2 Networks. High school correspondence courses and other related educational programs are also compiled for these services (see Table 4-3). For adult education, NHK broadcasts citizen's university

Table 4-3. Ratio of NHK's School Program Utilization (based on 1978 figures).

	Total Number of Schools	TV utilization rate	Radio utilization rate
Kindergarden	13,804	80.7%	19.6%
Primary School	24.697	95.1%	30.2%
Junior High School	10,593	52.8%	38.2%
Senior High School	4,681	54.9%	40.5%

courses. In addition, NHK compiles programs for specific audiences, such as language courses, vocational programs and programs for preschool children, and presents them at the times most convenient for the respective audiences.

NHK also pays careful attention to programs for the physically handicapped; its programs for the blind, the deaf and the mentally retarded are designed to take full advantage on the special qualities of radio and television. Especially, in a series of programs titled *Time for Deaf Viewers*, an interpreter himself presides over the program and uses not only hand signs but lip-reading, dactylologies and subtitles instead of putting the interpreter's hand signs in one corner of the screen.

All these educational programs are prepared in accord with long-range and systematic planning by the various production groups in charge. Accompanying textbooks are also available to supplement the NHK educational broadcasts.

Cultural Programs. NHK's cultural programs cover a wide range of subjects including history, art, international politics and trends of opinion, and new developments in science and technology, as well as practical information closely related to daily life.

Agriculture and fishery programs provide information on the latest farm and fishery management methods and the problems attending them. as well as on practical everyday farming and fishing.

A wide variety of business information from common problems in workshops to trends in the international economy are provided in industrial and economic programs. In addition, these present information useful to shopkeepers and those in small- and medium-size enterprises.

Medical and health programs present medical information necessary for the home, and pointers on caring for patients who need long-term medical treatment.

Programs for women include news commentaries for housewives, such as those on commodity prices and children's educational problems, as well as cooking, tea ceremony, and flower arrangement. There are also cultural programs on literature, art and science.

Such problems as environmental pollution or traffic accidents are dealt with from a scientific viewpoint in order to show how nature and technology should co-exist to promote a wholesome environment. Nature programs include those on wild life in Japan and overseas. One long-term series introduces seasonal changes in Japan by concentrating on flora and fauna throughout the country.

Entertainment Programs. NHK's entertainment programs are based on the goals of (1) bringing excellent entertainment arts into Japanese homes, thereby enriching the cultural life of the people, (2) preserving Japan's traditional entertainment arts and furthering their development, and (3) developing new artistic fields suitable for the distinctive characteristics of broadcasting.

Dramatic Programs. Among NHK's teleplays, the best known are a daily 15-minute serial depecting the courageous lives of ordinary people, produced on the basis of materials from everyday life, and a major year-long historical drama series presented every Sunday evening. The stars of these programs frequently become national idols, while the localities that have served as their settings often become tourist attractions. In addition, NHK's nonserial dramas, often written by Japan's noted playwrights, have won prizes at contests held in Japan and abroad. A number of Japan's major writers have started their careers with broadcast scripts, and they usually continue to write scripts even after they have become highly successful in the literary world.

Classic Arts Programs. The classic arts of Japan, such as the kabuki plays, non dramas and bunraku puppet shows, continue to be performed frequently in Japan today, but it is true that they have become less popular than before. NHK's program provide many opportunities for preserving such entertainment arts for future generations.

Light Entertainment Programs. NHK's established popularity is largely due to its well-produced entertainment programs as well as to its dramas. Quiz programs, light comedies and popular modern songs always draw a large audience. The most popular of such programs is the *Grand Musical Parade* televised on each New Year's Eve. This program annually enjoys ratings as high as 75 to 80%. Another program which has been a favorite for many years presents life styles and folksongs of various parts of Japan.

Music Programs. NHK's contribution in making a wide variety of music familiar throughout Japan since the start of radio transmission can hardly be overemphasized. Lovers of music ranging from Gregorian chants to electronic music and from Dixieland to rock may well be considered NHK lovers. NHK has done much to promote the work of contemporary Japanese composers by presenting and even commissioning their works. There is also an enthusiastic audience for the program series that presents folksongs of the world as on-the-spot recordings taken by NHK's own coverage teams, or supplied by overseas broadcasting organizations.

Special Programs. Just as joint research efforts are needed in the academic world in order to challenge complex problems, so the collaboration of experts in diverse fields is required in dealing with programs on major themes. NHK takes particular pride in its special programs created through the joint efforts of producers of dramatic programs, experts on news documentaries, and educational broadcasting specialists. These include *The Meiji Era*, a dramatized documentary series probing into the early stages of Japan's modernization; *Our World in the 1970s*, a series combining film clips and studio discussions to explore the problems of resources, human relations, education, environmental pollution; and *The Cultural Legacy for Future Generations*, a series of documentaries faithfully presenting relics of human achievements found throughout the world.

ENGINEERING AND FACILITIES

All NHK's broadcasting services, from their production through transmission (see Fig. 4-2) to final reception in every home in this country, are entirely supported by NHK's extensive engineering experience obtained in the broadcasting field over the past 50 years.

Production Techniques

Besides the fully equipped video production facilities at the broadcasting center in Tokyo, NHK maintains a number of color television studios and outside broadcast facilities at its seven regional headquarters and at many of the local stations.

Regular items of equipment for television production at a major local station include a studio with two color television cameras, a videotape recorder, a set of telecine equipment, outside broadcast van with two color television cameras, a videotape recorder and a microwave transmitter/receiver. The studio and van are also equipped with complete audio facilities.

Such equipment allows every NHK station to produce radio and television programs to meet local needs as well as to contribute to NHK's nationwide broadcasting networks.

Network Operations

The Technical Operation Center (TOC) at the broadcasting center is the hub of NHK's network control (Fig. 4-3). Daily program transmissions on the television, radio and FM networks from the broadcasting center are automatically controlled at TOC, supported by the computerized ABCS. Regular transmission of programs to the network according to a predetermined time schedule is the work of the transmission control room, where program monitoring and the emrgency operations are also carried out. In the transmission control room, along with the picture monitors, there are visual displays which indicate the programs, the item schedule, and program sources of each network. Besides the operation console, there are a number of videotape recorders and automatic audio tape players which play back the recorded programs to supply the network, under a control of a process computer.

One of the most remarkable functions of ABCS is network switching. Network switching of each program media, both video and audio, is executed by a group of special control signals superimposed on program signals. The switching system at every local NHK station is operated by these control signals received through the network.

Transmitting and Receiving Techniques

In the AM and FM radio broadcasting services, 173 transmitting stations for Radio 1, 141 for Radio 2 and 474 for VHF-FM together cover

Fig. 4-2. Aerial view of the NHK transmitting towers (courtesy NHK).

all of Japan (Fig. 4-4). Programs for medium-wave radio are mainly distributed by land lines to the transmitters. However, in the case of VHF-FM, stereophonic programs are distributed to major stations by means of recorded tapes and then relayed to smaller transmitters and translators by radio waves.

Television networks, the general and educational, consist of groups of main transmitters and translators, with an output power ranging from 50 kW in metropolitan areas to less than 0.1 kW in small, sparsely populated communities. The transmitters and translators total 2855 for the general and 2804 for the educational networks, respectively; all of these operate unattended (all figures as of Jan. 1979).

Due to characteristics of the signals in the VHF and UHF range, the service area of a television station is limited by terrain features. Providing all of mountainous and densely populated Japan with television service is the most formidable task to which NHK is obliged by law. NHK is constructing some 200 relay stations a year both for general and educational networks to extend its services to areas still having reception difficulties. For remote areas, NHK has developed television translators with solid-state circuitry, which assures high reliability and long-term durability. Also, in sparsely populated areas, NHK adopts the common antenna cable television receiving system (CATV).

NHK's engineering specialists deal with reception as well as transmission. In cooperation with electronics manufacturers and dealers, NHK tries to spread technical knowledge for better broadcast reception. NHK thus organizes technical courses to improve the skills of dealers and repair technicians throughout Japan. NHK is also dealing with reception difficulties in urban areas where high-rise buildings and numerous sources of electrical noise interfere.

INTERNATIONAL ACTIVITIES: INTERNATIONAL COOPERATION

NHK maintains close relationships with broadcasting organizations in various regions through its varied activities to promote the progress of broadcasting throughout the world.

By exchanging television and radio programs, NHK introduces Japan and her people to audiences in other countries, in order to advance mutual understanding, as well as to obtain programs by exchange to add variety to its domestic services. The steadily growing number of current issues and events of international importance has led to the increased exchange of news materials and programs via satellite transmission.

NHK is an active member of several international broadcasting organizations. Through mutual cooperation with their members, NHK endeavors to advance common interests and to promote interorganization relations. NHK has been a full member of the Asian Broadcasting Union (ABU) since this group was organized in 1964, and had helped prepare for its establishment since 1957.

NHK also participates in the technical cooperation programs sponsored by the Japanese government, and extends active assistance to the broadcasting organizations in developing countries. Specialists from NHK

Fig. 4-3. Partial view of the NHK control room (courtesy NHK).

are sent to such countries to help in constructing television and radio networks, equipping studios, giving guidance in program production and in training staff. NHK also offers training courses every year at its Central Training Institute for trainees from these countries.

JAPAN PRIZE CONTEST

In 1965, to commemorate the 49th anniversary of the start of radio broadcasting in Japan, an international competition known as The Japan Prize International Educational Program Contest was established to promote improved broadcasting. It is the first and the only international competition specifically for educational programs in radio and television.

The contest is for programs from school broadcasting and other educational series on all levels of primary, secondary and adult education, and it is open to all broadcasting stations and groups of broadcasting organizations from a country or a territory that is a member or associate member of the International Telecommunications Union (I. T. U.), which are authorized to operate a broadcasting service by the competent authority. The winning programs are selected by an international jury whose members are selected each year from among the participating organizations and noted scholars and specialists in educational mass communications.

RESEARCH AND SURVEYS

NHK has its own institutions for reasearch in the fields of culture and engineering related to broadcasting. These institutions carry out organized and systematic studies aimed at the enrichment of present and future broadcasting. The results achieved through such activities are used to improve the quality of NHK's own programs and to raise the standard of its broadcasting techniques, and are also made available to the public in line with NHK's policy of contributing to social, cultural and scientific progress.

RADIO AND TELEVISION CULTURE RESEARCH INSTITUTE

The Radio and Television Culture Research Institute of NHK was founded in 1946. A number of research projects are now in progress at this institute, including analysis of the use and effects of school broadcasts and the audience attitudes toward broadcasts. Furthermore, since NHK is trying to promote through its broadcasting the accurate use of spoken Japanese, the institute carries studies on the language and its correct usage for broadcasting with the collaboration of the broadcasting language committee.

The institute also gathers information on broadcasting both domestic and foreign in order to analyse current and future trend.

An exact analysis of social impact that broadcasting produces is another subject studied at the institute. The inherent characteristics of broadcasting and its social functions are analysed systematically.

With abundant historical data and materials, the compilation of the history of broadcasting in Japan is also in progress at the institute. The institute has its own special library on broadcasting where tapes and films of important events on prominent people and of vanishing folklore are collected and preserved. On the occasion of the 30th anniversary of NHK in 1956, a museum was set up at the institute, which was built on the site of the first radio studios ever operated in the country. This displays historical

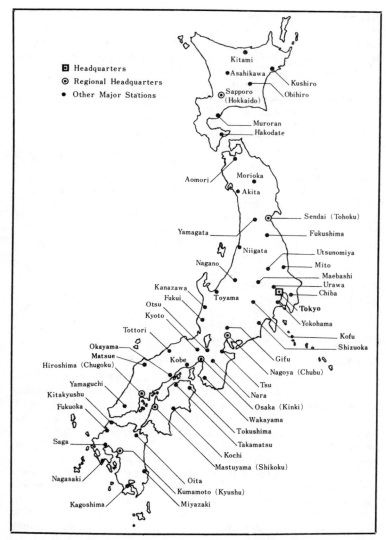

Fig. 4-4. Map showing locations of NHK's broadcasting stations (courtesy NHK).

materials showing the progress of broadcasting, and attracts some 80,000 visitors each year.

PUBLIC OPINION RESEARCH INSTITUTE

The nationwide time budget survey is conducted every five years by the Public Opinion Research Institute of NHK. The survey is made to analyze how individual Japanese spend the 24 hours of each day. The results become a fundamental factor in program compilation and other business planning of NHK, as well as essential data for various studies in sociology. Other systematic surveys by the institute include program ratings, public preferences regarding programs, and public opinions on important national issues, all by means of personal interviews. NHK is one of the very few broadcasting organizations undertaking such large scale surveys on a regular basis.

TECHNICAL RESEARCH LABORATORIES

Established in 1930, the NHK Technical Research Laboratories are active in every phase of broadcast engineering from production studios through broadcasting transmitters to receivers at home.

The laboratories carry out fundamental studies to improve current broadcasting techniques and to develop new technology for broadcasting in the future. Research is in progress on television pickup tubes, color television cameras, television tape recorders and other broadcasting equipment. Other areas of study include super-high-frequency techniques, acoustic engineering and information processing techniques.

In the field of future broadcasting technology, research has been made on still-picture broadcasting and on high-definition television. Broadcasting satellites are another subject studied at the Laboratories, in cooperation with the government plans to launch an experimental satellite. The developmental work on satellite-borne equipment and transmitting and receiving facilities has been making steady progress.

BROADCASTING SCIENCE RESEARCH LABORATORIES

The Broadcasting Science Research Laboratories conduct basic studies in aural-visual science and solid-state physics in order to develop future technology, in close cooperation with the Technical Research Laboratories. In the research into aural-visual science, psychological and technoloical analyses are made on the functions of the eyes and ears, and the results obtained have been used as basic data to improve the quality of existing broadcasting and to develop future braodcasting systems.

The principal effort in the studies on solid-state physics is toward developing new materials and components applicable to broadcasting. This includes experiments in the growth of crystals of various materials, detailed observations of crystal structures, studies on the magnetic, electrical and optical properties of magnetic substances, and work on quantum optics.

NEW DEVELOPMENTS

A number of new and recently undertaken developments will affect many phases of broadcasting.

Programs and Programming

In order to cope with changes in the listening and viewing modes of the audiences and with diversifying demands of the people upon broadcasting, NHK undertook a broad revision of the TV program compilation in April, 1978, and of radio program compilation in November, 1978. Special weight was placed on the revision of evening programs. Points considered in these revisions were as given below:

TV. The following are significant in the television area:

● Repletion of news and cultural programs. In addition to increasing the *NHK Special Features*, a large-scale news program presented between 8 and 9 p.m., to twice a week on GTV (general television service), new programs, such as the *Science Documentary, Reportage Nippon* and *Invitation to History* have been scheduled between 10 and 11 p.m.

● Establishment of the family hour. The evening time between 7:30 and 8 has been set aside as the family hour in order to cope with the listening and viewing modes of the people. Programs that can be enjoyed by the whole family are compiled for this period.

● Renovation of programs for children. In addition to scheduling a new program *600 This is Information Division* from Monday through Friday for providing fresh information for children every day, a new dramatic animation program has been scheduled.

● Repletion of local broadcasting. Efforts to enrich local broadcasting programs continued to be made.

● Enrichment of ETV (educational television service). Attractive cultural programs were compiled for the evening hours. Especially in order to cope with the scheduling of the family hour on GTV, such new programs as *Following up Topics, Interview Room, Show Era in Retrospect* and *My Autobiography* have been newly scheduled. These efforts were intended to attract viewers to ETV. The new program, moreover, have quite favorably been accepted by the audiences.

Radio. People listen to radio during the day while doing something else, but at nights they rather tune in to radio with a definite intent. For coping with such a situation, pertinent revision was undertaken over the entire scope of programming.

In the Radio 1 Network, wide programs presented in the morning were expanded and enriched. With housewives and proprietary business operation as the main target, programs that will provide useful information on everyday life, practical knowledge and higher level cultural information were compiled. Program compilation was arranged so that the listeners may listen daily to similar programs on weekdays in the same time zones.

In the FM service, classic musical programs and programs for younger generations presented on the most convenient hours have been expanded and enriched.

For weekends, enriched programs the listeners can enjoy at their best convenience were compiled for the Radio 1 Network and FM Service.

Test Presentation of Dual-Sound TV Broadcasting Commenced. From October, 1978, NHK commenced presentation of dual-sound TV broadcasting on GTV in Tokyo and Osaka. At present, two news programs a day and foreign dramatic films are presented in two languages. Some music and sports programs, such as *Masterpiece Album, Big Show* and the *Grande Musical Parade* presented at year end are broadcast under the stereophonic system.

Engineering

Experiments using the Medium-Scale Broadcasting Satellite for Experimental Purposes, *Yun*, which was launched on April 7, 1978, as a national project of Japan, have commenced.

Commencement of Experimental Satellite Broadcasting. NHK is in charge of experimental reception at the ground stations set up in various parts of the country. The experiment aims at obtaining basic data on the future induction of satellite broadcasting. Points under consideration following:

● Reception of high-quality TV signals in all parts of the country is to be confirmed.

● Receiving forms consonant with the conditions of location and environment in various places, such as cities, remote places and isolated islands, are to be studied.

● Attenuation of radio waves from the satellite on account of rain or snow is to be investigated.

● The performance and the operation of earth stations and receivers are to be studied.

● Data on the development of popular-type receivers in the future are to be collected.

● Various transmitting systems are to be tried out and the best system is to be selected.

As of November, 1978, the experiment so far is bringing expected results and much stock is placed on the results of experiments over the coming three years.

Frequency Change at Medium-Wave Radio Stations. On November 23, 1978, frequencies of 166 stations of the total 173 first network NHK stations and frequencies of all of the 141 second network stations were changed all at once. The change was based on the international agreement, under which the existing 10 kHz frequency separation was narrowed down to 9 kHz. As a result, almost all the stations in operation had to change their frequencies all at once.

Prior to the simultaneous change of frequencies, the second network station in Sapporo had its output power increased from 100 kW to 500 kW. The Sapporo station has thus become Japan's highest output station along with the second network stations in Akita and Kumamoto.

Development and Application of Soft Chroma Key. The chroma key is used widely for special effects in television. However, the screen synthesized using the chroma key can hardly escape the appearance of fitted-in pictures, since the blue of the chroma key back doubles up over some part of the picture; the boundary lines become blurred or finer parts disappear.

NHK has succeeded in developing the soft chroma key without the drawback attending ordinary chroma key. The newly developed soft chroma key is beginning to be applied, largely to the dramatic programs.

94

The use of this system easily allows synthesis of screens using transparent glass implements and cigarette smoke, which had been considered difficult to use in the synthesized screens. The shadows of persons reflected on the glass window can also be realistically synthesized.

Audience Relations

For an organization like NHK, the people's understanding and support is more important than anything else. Therefore, NHK is making every effort to heed the opinions and requests of viewers and keep them informed of NHK's intentions.

In July, 1977, an audience relations center was inaugurated in each of the local broadcasting stations throughout the country. At these centers, veteran staff members are assigned to quickly and accurately answer viewer's telephone calls and letters.

NHK also has the audience advisory council designed to listen to the views of the people on a national scale and gain their understanding of NHK's character and policy.

The council holds meetings three times a year at 53 places throughout the nation. A total of more than 1000 people of varied ages, occupations and groups attend these meetings and represent a small but solid cross section of the viewing public.

Moreover, NHK endeavors to further promote public understanding of NHK through direct contact with the people on such occasions as the collection of receiver's fees, program productions open to the public, news coverage and technical advice on better reception.

CHRONOLOGICAL HISTORY OF NHK

Mar. 1925: First radio broadcast by Tokyo broadcasting station, one of the predecessors of present NHK (March 22 has been set as Broadcast Day since 1934).

Oct. 1925: Japan's first outside broadcast (on-the-spot coverage of a parade).

Aug. 1926: NIPPON HOSO KYOKAI or Japan Broadcasting Corporation, a corporate juridical body was established to form a national broadcasting organization.

Aug. 1927: First on-the-spot broadcast of a baseball match.

Nov. 1928: First nationwide relay broadcast (on-the-spot broadcasting of the Enthronement Ceremony of the Emperor).

Feb. 1930: First successful long-distance shortwave relay broadcast from London.

June 1930: NHK Technical Research Laboratories was established, research on TV began.

Dec. 1930: First Japan-U.S. exchange of Christmas programs via shortwave.

Apr. 1931: The second radio network was put into operation.

July 1932: Deferred broadcasts of highlights of the 10th Olympic Games were relayed from Los Angeles.

Nov. 1932: Japan's first recorded broadcast beamed from Geneva.

Apr. 1935: Nationwide transmission of school broadcast began.

June 1935: Regular overseas broadcasts were inaugurated.

June 1936: On-the-spot broadcasts of the 11th Olympic Games in Berlin were successfully carried out.

May 1939: Experimental telecast was successfully conducted by Technical Research Laboratories.

Nov. 1941: First recorded relay broadcasting of a parliamentary session.

Aug. 1945: Broadcast of the recorded decree by H. I. M. the Emperor stating the end of the Pacific War.

June 1946: Radio and Television Culture Research Institute was established.

June 1949: Use of magnetic tape recorders began.

June 1950: Under the broadcast law, the corporate juridical body, Nippon Hoso Kyokai, turned to a special juridical body. Board of governors established.

Mar. 1952: NHK succeeded in trial wireless color telecast.

July 1952: On-the-spot broadcasts of the 15th Olympic Games in Helsinki.

Dec. 1952: First stereophonic broadcast using two radio networks.

Feb. 1953: NHK's Tokyo television station inaugurated regular television broadcasting.

June 1953: On-the-spot broadcast of the coronation ceremony of Queen Elizabeth II in a tie with BBC.

Oct. 1954: First practical application of kinescope recording in a TV relay of a Kabuki drama.

Mar. 1956: Broadcast museum opened in Radio and TV Culture Research Institute.

Nov. 1956: On-the-spot broadcast of the 16th Olympic Games in Melbourne.

July 1957: First Asian Broadcaster's Conference was held in Tokyo.

Dec. 1957: Experimental FM broadcast on VHF band inaugurated in Tokyo area.

July 1958: Use of videotape recorder began.

Jan. 1959: NHK inaugurated educational TV network in Tokyo area.

Apr. 1959: Special broadcast of the wedding ceremony of H. I. H. the Crown Prince.

Nov. 1959: Dispatch of NHK's special overseas coverage teams was started.

Aug. 1960: On-the-spot radio and TV broadcasts of the 17th Olympic games in Rome.

Sep. 1960: Regular color television broadcasts were commenced.

June 1961: Central Training Institute was established.

Apr. 1963: NHK Correspondence Senior High School was opened. Construction of NHK braodcasting center began.

Nov. 1963: First experimental transmission by the communications satellite was conducted successfully (between Japan and U.S.A., Europe).

Apr. 1964: The 2nd International Conference of Broadcasting Organizations on Sound and TV School Broadcasting was held in Tokyo by NHK.

Oct. 1964: Broadcasting of the 18th Olympic Games in Tokyo.

Oct. 1965: The first phase of construction of NHK broadcasting center was completed. NHK established "Japan Prize" International Educational Program Contest (annually). 2nd General Assembly of ABU was held in Tokyo by NHK as host organization.

Mar. 1966: Nationwide color television network completed.

June 1966: 1st ABU Administrative Council was held in Tokyo.

June 1967: First global live TV program *Our World* was broadcast jointly by 14 nations through four satellites.

June 1968: The 2nd phase of construction of the NHK broadcasting center was completed.

Oct. 1968: Color broadcast of the 19th Olympic Games in Mexico. NHK TOPICS started its operations.

Sep. 1969: First telecast of political view.

Jan. 1970: Experimental TV sound multiplex broadcast inaugurated.

Mar. 1970: Color broadcast of EXPO '70 in Osaka.

Jan. 1971: Experimental UHF broadcast inaugurated in Tokyo and Osaka.

Mar. 1971: Experimental SHF broadcast inaugurated at Technical Research Laboratories.

Aug. 1971: FM station in each prefectural capital throughout the country was completed.

Oct. 1971: All general television service programs were presented in color.

Feb. 1972: Broadcasting of the 11th Winter Olympic Games in Sapporo.

Sep. 1972: First satellite TV transmission between China and Japan.

June 1973: NHK Hall opened for service.

July 1973: The broadcasting center completed; concentration of NHK's functions accomplished.

Nov. 1974: 11th General Assembly of Asian Broadcasting Union was held at broadcasting center.

Mar. 1975: 50th Anniversary of inauguration of broadcasting in Japan.

Nov. 1975: International Broadcasting Symposium was held at broadcasting center.

Oct. 1977: All educational television service programs were presented in color.

July 1978: Experimental satellite broadcasting was commenced.

Oct. 1978: Test presentation of dual-sound TV broadcasting was commenced.

Nov. 1978: Frequencies at medium-wave radio stations were changed.

Chapter 5
Broadcasting in the
Republic of South Korea

Information Supplied by
Korea Research Institute of Journalism

The first radio broadcast in Korea was in 1927 and originated in Kyongsung from a 1 kW transmitter. Radio has developed from that time to the present. Now, radio was heard in almost all corners of the country. Some reception difficulties still exist but may soon be solved through the use of satellite relays. Confusion waves from North Korea also cause some reception problems.

Frequency regulations have been established through international cooperation. Less than one century since the discovery of radio waves, they have become highly useful in our daily lives in areas such as politics, economics, culture and industry.

AM RADIO

There are presently approximately 13 million AM radio receivers in the country. Some of the reasons for the large number of receivers (radio) as compared to television receivers are:

● No TV broadcasting during the daytime
● Low cost of radio receivers
● Radio is a domestic production.

The most effective to date year for radio broadcasting was 1978. It has been in conflict with television since TV came on the market. Some of the highlights for the year were:

● Establishment of a channel source for the comments of radio program specialization.
● Emphasizing news and information programs.
● Local citizen participation in the local network.
● Program development along with technical change.

The broadcasting networks are K.B.S., which owns 15 AM radio stations; M.B.C., which owns 20 AM radio stations; and T.B.C., which owns 3 AM radio stations. In 1979, there were 47 AM radio stations of which six were located in Seoul, the capital of Korea; 84.7% of all broadcasting stations are located in the countryside. These local stations depend on the Seoul station for 60 to 70% of their programs, while 30 to 40% of the programs are produced locally.

In the field of news broadcasts, radio is also competitive with television. Radio is developing more effective news programs during the hours when television is not broadcasting. The radio broadcasts must be rapidly relayed to the listener, because of the prearranged cooperative plan for the pooling of news relays which has been developed.

There are great amounts of documentaries and reports from overseas. Radio broadcasts are needed to analyze the factors which the listening audience is looking for. The program must be topical so it creates great interest in the listener. Because of the limited television broadcasts, radio becomes more challenging, especially during the early morning hours.

Another strong point of radio is the number of youths listening. Also, an increasing number of young married couples. Radio programming is broken down in the following manner: News, 20%; Education, 40%; and Recreation, 40%.

Programs for the housewife are broadcast in the morning hours and emphasize good personality. These programs feature famous actresses. Radio also gives traffic reports during the rush hours.

One of the great shocks to the youths was the termination of "Pop Song" programs during the night-time broadcasting. This was done because the broadcast stations felt the youth programs should be more educational by producing music such as national music, folk songs, and classical type music.

Great technical advances in automated computer broadcast systems have been made. AM stereo broadcasting will be imported to Korea. This is now in use in Europe and Japan.

FM RADIO

Because of the mid-wave reception problem, FM broadcasting helped to reach areas of difficult reception. FM has also expanded because of demands from listeners for stereo music and educational programs.

Programs such as *FM Concert, FM Festival* and *Classical Best 30* are some examples of our FM programming. The popularity of these programs gives some indication of the public acceptance of FM.

The broadcasting networks own the following FM stations: M.B.C., 2; T.B.C., 1; and K.B.S., none. There are a total of only six FM radio stations in the country as of 1979. Two of these are located in Seoul. FM radio receivers number approximately seven million.

The future looks bright for FM as it continues to expand not only its coverage but its programming also.

SHORTWAVE BROADCASTING

In 1956, the first shortwave broadcast was made for Koreans living abroad. It was expanded to 625 kW capacity in 1978 because of increased demands from the listener.

The K.B.S. is a representative of Nations for Overseas Broadcasting. Broadcasts are made in nine different languages:

Korean	11 times daily
Japanese	9 times daily
English	15 times daily
Chinese	5 times daily
Indonesian	5 times daily
Arabic	4 times daily
French	5 times daily
Spanish	6 times daily
Russian	4 times daily

TELEVISION

There are three television broadcasting networks in Korea: K.B.S., owned by the government; T.B.C., independently owned; and M.B.C., independently owned. One difference between the government-owned and the privately-owned systems is that revenue for government broadcasting comes from a fee charged the owners of television sets. Private broadcasting is financed through the sale of commercial time.

Television broadcasting has been developed along with the change and pace of living of society. In other words, the society influences the sensitivity of broadcasting.

As we know, the influences are from external (foreign countries) and internal (domestic) sources and require quality improvements. Under such circumstances, there has been improvement in organization and production of television broadcasting. The significant evidence is the increased sale of television receivers from three million in 1977 to five million in 1978.

The organization is planning to expand television broadcasting and encourage more independent programs. Program competition between the networks is encouraged also to give the viewer a better selection of programs. In 1979, there were 20 television stations on the air, of which three were in Seoul. K.B.S. had 9, M.B.C. owned 9, and T.B.C., none.

K.B.S. TV, the government-owned television system, has expanded from one local TV station to seven. Before this expansion, only private TV stations were operating in local areas.

Program content has been upgraded and more remote programs are being produced. Through the use of a satellite, more programs are reaching a wide area, where before it was only sports programs. Broadcasting from outside of the country has also expanded through the use of satellites.

In 1977, the Broadcasting Language Council adopted these guidelines:

- Use proper language
- Adapt foreign language properly
- Observe the proper use of grammar

- Avoid unnecessary jokes
- Eliminate meaningless language

In 1978, the council developed the sports broadcasting language as follows:

Soccer ....130 vocabularies
Boxing.. 77 vocabularies
Volleyball .... 53 vocabularies
Baseball ....187 vocabularies

All of these changes were established to standardize language usage.

The broadcasting language of foreign stations is a major subject to study in learning how to adapt to our own standard language with:

- Proper pronunciation
- Proper spelling
- Proper translation

Even under the different operational systems, good broadcast quality is desired, such as the work quality of the broadcasters themselves, good programs, and new technology. Therefore, there are continuous educational orientations and technical such as:

- Announcer seminars
- Producer seminars
- TV talent orientation
- Korean-Japan technical seminar
- NHK instructor seminar.

In order to expand the broadcasting technology from abroad, K.B.S. and M.B.C. have established advanced training in such foreign languages such as: English, Japanese and German. UNESCO and COLOMBO organizations are aiding in the study of broadcasting in foreign countries.

In management, needless to say, the personnel are very important. The broadcast business is a creative and selling business so it is very important to have excellent personnel.

Because of wage competition in the market, it is very difficult to obtain excellent new personnel. The trading companies are paying higher wages than broadcasting. New personnel hiring procedures are:

- Written test of our national language, as well as English and other foreign languages.
- Common knowledge of broadcasting.
- Thesis and recommendation from the dean of the school.

Technical improvements are a continuous project. Local television broadcasting stations have been established so more local programming can be offered instead of relying on key-station programs from Seoul. K.B.S. has installed television studios in the following cities to help provide this local programming:

| Pusan | Taeju | Chunchon |
| Taejon | Chunjon | |

BROADCASTING ETHICS COMMITTEE

The Korean Ethic Committee is strongly emphasizing the deliberation of a broadcasting ethic, in order to meet the social desire which contributes to the social educational function of broadcasting. In 1978, preliminary examination of content resulted in 353 rejections (13.4%), or a 2275 (86.6%) program acceptance out of a total of 2625. Some of the reasons for rejection were:

- Abusive advertising to children . . . (64 cases or 18.1%)
- Abusive language . . . (51 cases or 14.5%)
- Unpleasant expression to the emotional feeling . . . (50 cases or 14.2%)
- Exclusive expression . . . (44 cases or 12.5%)
- Long hair . . . (30 cases or 8.5%)
- Excessive body exposure . . . (18 cases or 5.1%)
- Others . . . (33 cases or 9.3%)

This examination of advertising is conducted every year. The standard ethic is not only for consumer protection but also the change of public life consciousness. These preliminary examinations are done prior to being telecast. The reason for this prebroadcast examination is the fact that advertising is really penetrating the countryside and is so influential.

The ethic committee has recommended that programs should not influence the public morality in the following manner:

- Restriction of long hair on the screen
- Improper life style (discrimination between rich and poor)
- Excess exposure of the nude body
- Dress style
- Physical body contact of the sexes.
- Improper romantic series.

There has been a significant reduction of ethic violation through broadcast control. The violation areas are: New violation, 12.8% Social education violation, 11.3% Entertainment violation, 48.6% Commercial violation, 27.3%.

PROGRAMMING

The emphasis in the area of television programming has been that of quality. More educational programs are needed, as well as programs on Korean culture, history, music, etc. The golden hour and family hour are where these quality programs are needed. One problem is the limited time allocation of 5½ hours of broadcasting on week days. It is a major goal to develop a full-scale of programs instead of short series.

Drama

Demand from the viewers has brought about a need to develop full-scale television dramas. Manpower, materials, production techniques, and title selections are needed for future success.

The drama is probably the most economical type of program because it can be divided into daily program series. This type of programming creates more curiosity because of its continuous nature.

Weekend drama is improving significantly. K.B.S. has put a great amount of time and effort into experimentation for the improvement of these programs.

With the influence of *Roots*, imported from the United States, the drama has improved dramatically. Due to a great deal of unsatisfactory complaints about the TV drama, the three TV stations have worked hard to make the programs more competitive. Even with the improvement of TV drama, a lot of difficulties still remain.

Children's TV drama is one of the weakest types of programming because of the priority of drama development for the adult program which comes ahead of children. It is time to see the plan for drama develop for children in the future.

Foreign programs are imported from other countries. These programs are broadcast on domestic television during the following hours:

- 30 minutes broadcasting at 6 p.m. for the children
- 60 minutes broadcasting at 10:30 p.m. for the adults
- Movies at 10:30 p.m. on Saturday and Sunday for the adults.

There are some questions whether these foreign movies should be broadcast at the golden hour, but it is not easy to use the golden hour for foreign films.

The time has arrived to admit more domestic films to the TV screen. Of course, there are many problems involved, techniques, crews, actors, and actresses.

It was a great shock to the foreign movie fan in 1978 because the government suspended the broadcast of foreign films on weekends at midnight. Foreign films and programs are still seen on domestic television, however. Some of the more popular programs are: *Little House on the Prairie, Bionic Woman, Six Million Dollar Man, Rich Man, Poor Man, Captain and Kings, Once an Eagle, Executive Suite, Aspen.*

The BBC in England and Time-Life in the U.S.A. produced cooperatively a documentary drama entitled *Fight Against Slavery*. It was broadcast at 10:30 p.m. Monday, Tuesday and Wednesday nights by M.B.C. and was very successful.

In an effort to develop musical programs there was an "International Music Festival" With ten countries involved held in Seoul in 1978. It was very successful as a first celebration.

M.B.C. had its second college Musical Festival. It was a great opportunity for the college students and helped with improvement of musical programs.

Selection of materials from overseas is done carefully because of:

- Cultural differences
- Custom background
- Actors and actresses unfamiliar to the viewer.

Under the tight budget situation we have to consider educational materials most compatible with our own customs, otherwise it is wasted.

The time has arrived to admit more domestic films to the TV screen. Of course there are many problems involved: technique, actors and actresses, financial, etc. Either more domestically-produced film for TV or at least the joint production with European or American countries should be tried.

News and Sports

Some broadcasting stations have opened their own news centers in order to broadcast more live news to their viewers. Due to budget insufficiencies, the use of satellites is not used as often as desired. Most news is coming from VISNEWS or TV news supplied through air mail. The Kromaki News system is also been very successful for the news. This system takes a great deal of art design along with the report from the journalist.

In sports the Korean Women's volleyball team played in Moscow in 1978 and the games were broadcast through the international telephone system. The Asian Games were also broadcast live from Thailand.

The depth of television news time is still limited, even if the average individual income is over $12,000. In other words, the news time is still needed to develop more depth, for even the national situation has been progressing both politically and economically.

Along with economic growth, there was apparent feeling that the consciousness structure of society is still paralyzed. It was very sad that we could not broadcast fully to the public because of news restrictions and regulations. The public is demanding up-to-date news. This requires a lot of effort from the broadcasters. Although some news is broadcast, the public wants complete news of all subjects.

One of the great news broadcasts was presented live from Russia by the Korean Broadcasting System when a KAL airplane landed in a Communist country on April 21, 1978. There is need for more TV broadcasting from other nations.

There is need of programs concerning social problems such as:

- External school work for the pupils.
- City traffic problems
- Housing
- Pollution
- Protection of Nature

Female news announcers have been hired since 1977. This male and female double cast system works very well. It will require a great deal of study to produce good quality female announcers in the future.

Comedy

The history of comedy shows was one of slap-stick humor as it was on the stage. K.B.S., T.B.C., and M.B.C. have tried to bring satisfactory comedy shows to the viewers. Unfortunately, the government ordered comedy shows to be suspended in 1977. The major reason behind this was the influence (bad) on the educational programs. Entertainment and recreational programs, especially the comedy programs, were becoming lower in quality. However, because of viewers protests, the comedy shows are gradually coming back.

Advertising

One of the major reasons broadcasting has been growing successfully in Korea has been its economical growth. The major financial revenue come from commercial advertisements. According to national law, there is a limitation on the number of times advertisements can be broadcast. Another revenue law is that 10% of a commercial fee is taxable.

Advertisement is an industry of information, and so the industry is composed of highly educated personnel. Because the broadcasting business is so highly sensitive as an information industry, new ideas are always needed. They should be setting aside sufficient capital for the future development of color television operations.

There was quite a change in commercial advertisement policy along with new production manufacturing equipment imported for quality improvement to domestic products by the free import policy. Compared to previous years, there was an increase in GNP of 133.6%, a 145.2% increase in total advertising, and 146.6% increase in broadcasting advertising. These figures show that broadcast advertising increased more than the GNP in 1978. For eight years, from 1970 to 1978, the GNP grew an average of 28.4%, while advertising increased an average of 34.4%. This indicates a 6% a year average in favor of advertising.

Broadcast advertisement fee increases are agreed upon between the station and the program sponsor. Fees have increased 34% for television and 46% for radio. Even with these increases, there are still financial difficulties because of:

- actor/actress payments
- depreciation of equipment
- program content expenditure
- program profit.

According to research, broadcast advertising is more effective than newspaper or magazine advertising. The top commodity of the advertisement business is food.

Broadcast advertisement has changed drastically along with social and economic change. This means that broadcast advertisement is linked to the public life and it functions so sensitively. As a long-term plan, it is hoped to see the advertisements on television in color.

THE FUTURE

Through the use of satellite broadcasting, it is hoped that we will have better international understanding and cooperation.

In order to improve the communication between viewer and government, there will be more channels in the future. In 1908, the following subjects are the goals:

- Color television
- Technical revolution
- Image aesthetics.

Based on the investigations as of December 31, 1978, there were 5,135,496 TV sets in operation. This figure indicates that there is a great market for more and better programs, more stations, etc.

In a few years, there is a great hope to operate color TV broadcasting. K.B.S., M.B.C. and T.B.C. have almost-installed color TV broadcasting systems but these require high operating costs. Therefore, the advertisement fee will be increased along with operating costs, which is the major financial source. For this purpose, the management is required to develop selling power which means program development with efficient management as the key factor.

Electronic newsgathering is being tried, in expectation broadcasting of the ENG age in the near future.

Because of the lack of VHF-TV channels, we will be forced to add additional UHF-TV channels.

It is hoped to receive channels through satellite transmission if the completion of technology preparation is achieved and sufficient financial funds are available.

As we have seen, there has been a rapid development of broadcasting technology in such a short period of time by the domestic electronic industry. There is a possibility in the future for great development in the broadcasting field.

Chapter 6
Broadcasting in Israel

Ari Avnerre
Israel Broadcasting Authority

Radio was a relative latecomer to the Holy Land. It was not until early in 1936 that public broadcasting was inaugurated by the then British Mandatory Governemnt of Palestine. It was only a few weeks after the opening ceremony that the country was engulfed in a wave of Arab terrorism (1936-1939), adding to the seething problems of a newfangled broadcasting operation the precarious balancing act needed to maintain credibility in the midst of a war-like situation, with everybody listening in and comparing notes.

These conditions came to an end only to make place for a real full-fledged war—World War II. Paradoxically, it made the broadcasters' role easier as far as credibility among the Jewish population was concerned: wartime censorship notwithstanding, the overall aims and ideology were such that one could identify with it without undue difficulty. For understandable reasons, Axis power broadcasts had no progaganda impact whatsoever on the Jewish population. The very term—propaganda—acquired its pejorative connotation thanks to Goebbelsian usage, and was thus never attributed to anybody or anything siding with the Allies.

PALESTINE BROADCASTING SERVICE

As World War II drew to a close, with Allied victory no longer in doubt, the struggle for the future of the country began anew, and the plainly partisan PBS (Palestine Broadcasting Service) reincurred all the stigmata of an unabashed government mouthpiece. No attempt was even made to differentiate reporting from propaganda. The man in charge of the news operation of the PBS himself broadcast weekly talks plainly calculated to put across the British administration's point of view. Rival views were never aired.

Oddly enough, the BBC tradition of fairness was little or no help, for in addition to its proud independence, that tradition had a streak of

Reithian paternalism; and what they could do, when superimposed on a colonial administrations expediency calculations, can all too easily be damaged. Such phrases as "we do not go into that kind of matter" were a common utterance in the PBS, and "that kind of matter" did not necessarily refer to current social taboos but to a large part of what constituted even then the bread and butter of journalism.

CLANDESTINE ISRAELI BROADCASTING

With World War II over, holds were no longer barred. The early battles of what has proven to be the curtain-raiser to Israel's War of Independence were carried into the air. It was the early postwar years tnat saw what proved to be the "other parent of broadcasting" in Israel— clandestine underground broadcasts, carried on by Jewish paramilitary organizations, the Hagana (defense), the clandestine forerunner of Israel's armed forces, as well as two smaller groups bent on the use of force to achieve a political change and establish a Jewish State—the Irgun Zuvi Leumi (National Military Organization) and Lohamey Heruth Yisrael (Fighters for the Freedom of Israel).

It was the Hagana Broadcasting organization to which some founders of broadcasting in Israel could proudly trace their antecedents, by direct lineage, in addition to the PBS. Not that there was a sharp line of division: Quite a number of veterans had a foot in both camps, the present writer included, referring to ourselves as Dr. Jekyll and Mr. Hyde, and never doubting who Dr. Jekyll was.

ISRAEL BROADCASTING SERVICE

Though sorely lacking the attibutes of an orderly transfer of power, the termination of the British mandate and the emergence of the State of Israel neither stopped broadcasting for even one day, nor was the basic structure of broadcasting, as part of the administration, changed over night. With some hastily recruited reinforcements, the Jewish staff of the PBS and some Hagana broadcasting veterans carried on operations in makeshift premises before they actually knew where their salaries were going to come from.

We need not have worried; nor were these worries paramount, with shells booming and shrapnel whistling around our ears in besieged Jerusalem. In a matter of days, broadcasters found themselves part of the government information services, run, with such other units as the public information office, by the ministry of the interior, not in a small measure because of the personal whims of Israel's first minister of the interior. He didn't have much use for cops, so a separate ministry of the police was formed. But information he liked, so it wound up in his ministry.

With Government organization assuming more of a semblance of rationality, the information services were, after a short while, bodily transferred to the prime minister's office, which acted as a catchall for those government departments for which it was not easy to assign definite ministerial responsibility elsewhere. The idea of a separate ministry of information still carried with it totalitarian overtones, and was therefore not contemplated. A lengthy process, sometimes painful, sometimes bordering on the ludicrous, began. Mutual education, easing broadcasters into their role within a democratic society, and informing political leaders what broadcasting was all about, even technically speaking.

On the first day of the existence of the State of Israel, as well-known Jewish political leader was to be the first to address a national audience on the air. The broadcast was scheduled for 6 p.m. An overgrown reception committee was awaiting his arrival at the gates of our make shift studios, in a girl's home in Jerusalem. There was no sign of him, either by 5:50 or by 5:55, and at 6 p.m. the announcer went on the air to declare, in a sepulchral voice, that "because of unforeseen difficulties" he was going to play a few records. At 6:05, the gawky figure of the speaker made its appearance at the by then deserted gates. When the situation was explained to him, he could only blurt out, "B-b-but I'm only five minutes late!"

Nor was all broadcasting in the country incorporated into the "Israel Broadcasting Service," as it was then. An exaggerated notion of propriety dictated the external broadcasting services to be established under the aegis of the Jewish Agency, which as a private organization, the argument ran, need not concern itself with the propriety or otherwise of one state addressing itself to the citizens of other sovereign states, over the heads of their governments, as it were. It took complete disregard for this fine point by foreign monitoring services, as well as a certain amount of administrative waste and mutual toe-tredding, to convince everybody that

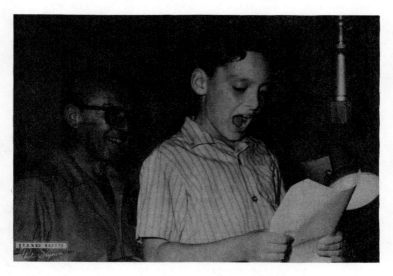

Fig. 6-1. Children's hour (radio) veteran actor A. Ben Yosef and a young participant (1969).

the duplication of, say, record libraries had no political significance. "The Voice of Zion for the Diaspora" was then incorporated into the IBS, as its external service. Another outgrowth of those early years survives in the form of a separate army radio station (not subject, for the purposes of programming, to the Israel Broadcasting Authority).

Anxious to maintain democratic precepts of broadcasting even in conditions of a monopoly, successive governments resisted temptation to make the service into their own mouthpiece; and the legal seal of approval on this state of affairs was put in 1965, when the Broadcasting Authority Law was enacted by the Knesset (parliament), in effect divorcing broadcasting from the actual administration with its concomitant party political complexion. The authority was not, however, meant to keep aloof of the body politic as such. It is required by law "to provide the broadcasts as a national service" and "ensure that the broadcasts . . . enable appropriate expression of different outlooks and opinions prevailing among the public and supply reliable information."

The law was substantially amended only once, at the end of 1968, to provide for the inclusion of television; but more about that later. The basic provisions of the law remain virtially unchanged. Two public bodies were established: a plenum and a "managing committee" (roughly a board of governors) to run the authority (now with 31 members and seven members respectively). The members of both bodies are in effect government-appointed, but political convention has established two practices safeguarding their independence from the government of the day: members of the public bodies are in practice selected so as to represent a wide party political spectrum, covering all the important parties in parliament, and once appointed, members are no longer subject to instructions either from the government or from any other outside source in the exercise of their duties. The law itself limits the number of state employees allowed to serve on these bodies, no more than four on the plenum and no more than two on the board of governors (and there is none in the present board of directors). Nor was actual independence allowed to remain a dead letter: the cause celebre in this context was the occasion, later on, when the authority put television broadcasts on a full-time basis, including the Sabbath, in overt defiance of the government's wishes.

The above mentioned reluctance of successive governments to make broadcasting the handmaiden of administration had its counterpart in the growth and self-assertion of professional traditions among broadcasters themselves. There were thus no days of sudden leaps to freedom when the Broadcasting Authority Law was enacted. Past traditions could be carried on with a now legitimate disregard for red tape and even previously futile weight throwing.

TELEVISION

The next turning point in the history of broadcasting in Israel came in the wake of the 1967 Six Days War, with the establishment of television. Much reluctance and downright opposition had to be overcome. In addition to a snooty disdain of the supposed cultural debasement inherent in such a mass medium, the fears of the press and the like, treasury economists were terrified by the prospect of the colossal expenditure in foreign currency. The time was somehow never right for that; it still isn't. Several developments, however, made the politicians bow to the inevitable. In 1966, instructional television for schools was started by the Rothchild Foundation, under the protective wing of the ministry of education. Education has always been something of a fetish among Jews, so nobody took a close look in that particular gift horse's mouth and it did not encounter serious opposition.

In the surrounding countries, television, as a means of government guidance, occupied a fairly high place on the list of political priorities. The dreaded drain on foreign currency reserves started taking place anyhow, with people buying sets to watch ninth grade math on instructional television and soap operas on the Jordanian channel or what have you.

Fig. 6-2. A well-known actor, A. Lebiush, hamming it up in front of the premises.

Both government and parliament gradually faced the situation where the question was no longer whether the people were going to watch television, but rather what they were going to see on their screens. The 1967 war provided a raison d'etat catalyst: the supposed impact of Arab television stations on the minds of the Arab population, of Israel itself and of the newly-acquired territories, dictated that they should be offered an alternative. Under the prime minister's office, a team headed by a communications professor was charged with the task of starting emergency Television Broadcasts with special reference to broadcasts in Arabic. Time, and a brief span of it at that, did away with the well-meant misconception, that television broadcasts in the country could be based on a minority language. Almost from the beginning, programs intended for the Jewish population, produced or subtitled in Hebrew, became dominant.

Discontent with the reality of television elsewhere put before theorists the temptation to disregard other peoples' experience, start everything from scratch and arrive at an entirely new thing. This illusion fared no better. Eventually, general considerations pertaining to the place of broadcasting in a democracy, as well as administrative despair of the ability of a small government office tail to wag a huge television dog, resulted in the inclusion of television into the broadcasting authority, even before it started operations on a regular daily basis.

Though instructional television on one hand and some hesitant preparations of the authority on the other have laid some foundation for professional television, the suddenness of the "emergency" decision to start operations has in effect catapulted everybody into a new communications era without adequate preparation, let alone organic growth. People with experience in television abroad, a motley crowd with extremely diverse professional traditions, were hastily pressed into service, most of them temporarily, to forge the first link. Zealous amateurs took their time becoming professionals themselves, and did so at viewers' expense.

The authority often found itself publicly defending sheer inexperience, in order to safeguard professional independence. Sights were inevitably set too high, well beyond capacities and the means available. Ambitious programs were done on a shoestring—and looked it. Directors of television euphoriously came in and shortly left in black despair. In a matter of a few years, however, stability set in, professionalism asserted itself and programming became realistic.

In recent years about half the television output in Israel is locally produced. Audience research has proved that not even the most popular shows acquired abroad could compete in ratings with local fare. At present, a locally produced consumer-oriented program heads the popularity list with 72.1% of the adult population watching it every week. Given the special sensitivity of the public in Israel to news, the main news bulletin maintains a rating higher than 60% every day.

Television still operates in Israel under well-nigh impossible constraints. Its channel is shared with instructional television, which means that there is no program for the general public until 5:30 p.m. What is considered peak viewing time is in part devoted to broadcasts in Arabic (from 6:30 p.m. to 8 p.m.). Means, need it be said, are sorely inadequate.

In the main, television is financed by license fees, which cannot be raised out of reach of a working family. Since the number of households in Israel is somewhere within the range of three-quarters of a million, it means that the sum total cannot reach anything approaching the adequate. Extension of services into a second channel, therefore, implies another source of finance.

MODERN RADIO BROADCASTING

It was only recently that radio stopped reeling under the impact of the blow of the introduction of television. Early 1979 audience research figures for the first time showed radio was rallying back in holding the attention of the audience, with a concomitant zero growth of television audience.

Not that radio was ever relegated to the limbo of insignificance in Israel. Its comprehensive news service—especially in times of war, when the station in effect became all news with intervals for the rest—was sufficient to hold public attention. During the 1973 war, more than two-thirds of the civilian population listened to more than 10 hourly news bulletins every day. Still, radio took its time to adjust to changed listening patterns, with once peak-time audience severely curtailed by television.

Radio broadcasts in Israel now go out on six channels: *Channel One* broadcasts serious music, some drama, highbrow talk and a large number

Fig. 6-3. Arabic folk tale televised in the original language ("The Happy Cobbler").

of minority-oriented programs, including foreign-language broadcasts for new immigrants as well as special programs on say, agriculture or chess. *Channel Two* has news as its backbone, with 20 news bulletins a day, three comprehensive news magazines and a number of special interest news shows such as sports, the economy, etc. It also has a wide variety of light entertainment and talk shows of wide appeal.

Channel Three is mainly light music interspersed with hourly news. *Channel Four* broadcasts mainly in Arabic with a number of other foreign-language news bulletins. *Channel Five* is the external service channel, broadcasting in 14 languages. *Channel Six* is the army radio, which has its "civilian" programming controlled by the Broadcasting Authority, with its main orientation on a young audience (its news bulletins are "feeds" from Channel Two).

The impact of broadcasting on society has evolved over the years with attitudes determined both by the history of broadcasting itself and by the development of society at large.

A curious ambivalence developed in public attitude to broadcasting in the '30s. On the one hand there was pride in broadcasting as a new conquest of renascent Hebrew language and culture. Top authorities in their fields were prepared to chip in, for a ludicrous fee, and lend prestige to broadcasting as well as bask in its light. On the other hand, the attitude to broadcasting as a medium of information, due to its being run by a colonial administration, could best be summarized in an almost 40-year-old cartoon: it showed a man taking his radio to a repair shop, with the caption reading "Can you do something about it, it keeps lying?"

No more could the clandestine stations be considered a useful corrective, in a professional sense. They did evoke identification and trust, but as far as standards of dispassionate reporting were concerned, they were patently out to carry a partisan message rather than merely inform, let alone entertain. A characteristic report describing marauders who had come to a sticky end was rounded off with a quotation from Judges 5.31: "So let all thine enemies perish, O Lord."

The early days of the State of Israel, with the declaration of independence broadcast live, were a honeymoon between broadcasters and the public. "Our" state could do no wrong—broadcasting or otherwise. However, life in Israel has always been high on party political content. As patterns in this respect were crystallized, sensitivities and suspicion set in. It has earlier been said that successive governments have refrained from abuse of the medium; but the very fact that it was controlled by them produced reservations, at least in that part of the public that was in disagreement with the government of the day.

Due credit should be given to Israel's first prime minister, David Ben Gurion, who had thrown his considerable authority behind a code of broadcasting ethics that placed truthfulness and fairness above his own political needs. He gave short shrift to members of his own party coming to him with a "how dare they" criticism of unpalatable broadcasts. Unfortunately, he went as far as leaning over backwards, ordering the news room (the only case of its kind) to "kill" a story carrying his *own* retort to a bitter

Fig. 6-4. The Bard visiting the miraculously survived aging Romeo and Juliet. From E. Kishon's sequel to "Romeo and Juliet" called "Oh, Julia!" l-r: A. Lavie, S. Shani, Y. Yadin.

political enemy, whose tirade itself previously had been carried by the station: "Bless you, nobody is going to accuse *him* of making political use of the broadcasting service!"

Mythogenic attitudes were not unknown even among broadcasters themselves, who imagined the dead hand of governmental censorship whenever a senior editor refused to go along with young staff members' flights of fancy. The unhealthy situation of a discrepancy between broadcasters' professional ethos and their legal status as civil servants came to an end with the establishment of the Broadcasting Authority, inevitably raising exaggerated hopes in the hearts of opposition politicians: they seemed to assume that the impartiality of broadcasters could somehow do away with the predominance of government in its capacity as newsmaker. Ironically, when a change of government did occur, one heard similar accents in the ranks of the newly formed opposition.

Politicians sometimes took the wording of the law: "enable appropriate expression to different outlooks and opinions prevailing among the public" to improbably mean that air time should somehow be apportioned to political parties according to parliamentary strength, even where hard news was concerned. Some politicians invested in stop watches, and at least one of them taught his young son to operate it and keep a close check. Politicians apart, the public gradually accepted broadcasting for what it was, and sometimes, engagingly though uncomfortably, attributed to it all but divine veracity.

MODERN TELEVISION

With the establishment of general television broadcasts, patterns of behavior basically changed. Already in its early days, phrases like "see you on Friday night, after Forsythe" became common currency. The audience ratings, mentioned elsewhere, could not have been possible except in a society that is television-oriented to a fault. Early 1979 audience research figures showed, for the first time, a slight drop in television audiences. But it is still premature to conclude that finally the love affair between the Israeli and his television set is beginning to wane. The drop is there only in comparision to the peak of early 1978 and late 1977, when President Sadat's visit and its aftermath shot audience figures into an all time high. There is no drop off in television audiences as compared with 1977.

Born of one war, Israel television came of age in another. The 1973 war was the first one that could be televised, thus making Israeli's closely knit society even more closely identified with the men and women engaged in the actual fighting.

The subsequent period of national soul searching, ultimately perhaps resulting in a party political upheaval in parliament, had its beginnings in the mood created during that war and in its wake. The temptation to compare the impact of Israel television in 1973 to that of American television in the Vietnam war should, however, be resisted. This was not a war fought in a faraway place to defend hard-to-define national interests, but a war of survival, pure and simple. Therefore, the result was not a growing feeling of "let's get the hell out of there," under the impact of the ugly realities of war brought home, but rather a mood of national introspection bringing about divergent answers to all important questions and upsetting traditional socio-political patterns.

In another sense, television has turned Israel into part of the "Universal Village," contracting distances and conducive to shared experiences with the rest of mankind, or at least its western part. Television may have done for certain kinds of pop fare what radio did for music in the previous stage. Surprisingly, audience ratings, considered to be fantastically high by international standards, were also commanded by such "elitist" programs as a Mozart opera or a Shakespearean play. This might be due, to a certain extent, to indiscriminate viewing, a very mixed blessing. But it resulted in untold cultural riches being brought to a mass audience, which is an unmitigated blessing.

THE FUTURE OF BROADCASTING

Though, in the opinion of the present writer, the independence of broadcasting is no longer a problem, it remains an issue. In the final analysis, whether or not broadcasting loses its functional independence in a democracy depends on the viability of democracy in society as such. Suppressing news on the electronic mass media is, after all, of little avail unless accompanied by a muzzling of the press and ultimately word-of-mouth rumor as well. Since there are no signs of the likelihood of such a development in Israel's society, the bark of the watchdogs will always remain too loud for an effective Gleichschaltung of the media. Nor does anyone in power seem to be unprincipled enough to want it or foolhardy enough to attempt it.

An old Israeli joke describes a man coming out of the broadcasting station, being asked what he was doing there, and saying "I offered mmmm-my ccccc-candidature as an aaaaaa-anouncer"; and to the inevitable question whether he got the job, replying "I dddd-don't bbbb-belong to the right ppp-party."

Such attitudes as here ridiculed, however, do persist, and party political lobbying for senior appointments—however well meant and however futile—do nothing to allay such suspicion. However inconvenient such a state of mind and affairs could prove to the board of directors, even this has its useful function as a constant reminder that the national and nonpartisan standards of broadcasting have to be maintained.

Among the problems facing broadcasting in Israel in the future, the question of the means available to it is paramount. Radio went a long way in solving its financial problems by allowing limited advertising on some of its channels. It is only thanks to advertising the radio was able to offer a rich and varied menu to its audience.

Television is still barred from accepting advertising and its development is unthinkable without it. The plain fact is, that if television is to develop and branch out into one additional channel at least, the only alternative to advertising is a sizable subsidy from the government. This is both unrealistic and undesirable. Unrealistic, because the order of priorities in Israel's society and government does not place a very high premium on the needs of communications. The dire necessity of

Fig. 6-5. Scene from a TV children's hour, a comic classic by Sholem Aleichem.

maintaining military strength, the kind of cultural and social conscience refusing to contemplate significant infringement on such things as education, health and welfare, combined with an economic imperative to keep a tight lid on the national budget as a whole, pits subsidizing broadcasting very near the end of the queue. Nor is it desirable to make the Broadcasting Authority Law sound hollow, by having the treasury paying the piper, call the broadcast tune.

Arguments put forward against the introduction of advertising on television are in part reminiscent of those raised against advertising on radio (fear of competition by the press, premonitions of increased consumption when the nation needs to tighten its belt) and in part—of those raised against the introduction of television itself (debasement of culture, pernicious influence of the young etc.). Since introduction of television advertising probably needs parliamentary sanction, it really depends on recognition by the politicians' community, that better and more varied broadcasting may be as creditable as, say, bowing to the press.

In fact a government-appointed commission has already come out with a report advocating the establishment of an additional television channel, supervised by a public body but operated by private interests financed by advertising. As against this, two kinds of criticism of the commission's findings are often voiced. One is the old argument of this not being the right time for the national expenditure involved. Another is apprehension that a second channel operating outside the Broadcasting Authority would result in competition for mass audience only, and thus bring about an overall lowering of standard rather than develop mutually complementary outlets.

Since the size of the country and the strength of its economy would not bear an American-style private enterprise multi-channel approach, public monopoly of broadcasting can be more rational when it has a united purpose, to give the maximum service to the largest number of people, without neglecting various minorities. In any case, fear of an overpowering monopoly will, within a matter of a few years, be alleviated by the second generation of satellite broadcasting, making foreign television broadcasts available to local audiences.

Israel television is still black and white. It is only partly equipped for color and color is "erased" from foreign programs to ensure uniformity of output. Still, color sets are being imported and sold at a quick rate and here again foreign broadcasts may force Israel's hand: those who have color sets get Jordan in color and make local fare look drab by comparison. Engineering ingenuity has also produced a relatively cheap device known as the "anti-eraser," enabling owners of color sets to restore color to foreign programs broadcast on Israel television. This last development has made a mockery out of the very process of erasing, and the practice may be discontinued in the near future, thus forcing television to accelerate its own capacity for production in color, so as not to make local programs automatically inferior to imported ones. There is some political opposition to all color—puritanically motivated—as well as the perennial

Fig. 6-6. A Nazi war criminal making haste when spotted by an Israeli TV team in Europe.

anguish of treasury economists, lest we send the nation on a spending spree.

With radio coming back into its own, two major lines of development are envisaged. One is the growing importance of FM channels, with AM bands in the Middle East getting more and more crowded by oil-financed powerful stations. Another trend could possibly be the development of urban local radio stations. This latter possibility is still clouded by a very big question mark: the hometown newspaper has never been a success in Israel, with the exception of trashy throwaway sheets wholly financed by advertising. Israel has tended to act and react as one single community, and it is still doubtful whether local broadcasting will be able to marshall a sufficient following to justify advertising and survive.

A word about external broadcasts. In the main, they are directed to Jews all over the world. For some of them, living in open societies, these broadcasts are no more than a form of direct contact, a way of getting more quickly up to date. For other Jewish communities such broadcasts may be the only means of giving them any information not only about present-day Israel, but also about Jewish history and heritage, with no corrollary services available in the form of newspapers, periodicals, freely accessible libraries and the like. This often has to be done in the face of malicious jamming.

Israel broadcasts in Arabic have, over the years, built a reputation of high professional standards and credibility. In time, they may become a bridgehead of a new mutual awareness in a peaceful region.

Chapter 7
Broadcasting in Canada

Albert Aber Shea
Canadian Radio-Television & Telecommunications Commission

Guglielmo Marconi came to the Atlantic Coast of Canada to successfully receive the first transatlantic wireless signal from England.

He established a voice broadcasting station in Montreal in 1919, Canada's first. Its successor still survives as CFCF-AM, Montreal. Also, Marconi established a Canadian manufacturing company which is still a going concern.

Radio stations sprouted like mushrooms, mainly in the principal cities; in the 1920s more than 60 private stations came into operation.

In 1929, the first Royal Commission on Broadcasting reported. It recommended a public service corporation to serve all Canadians with Canadian programs, and proposed a role for the provinces in the conduct of broadcasting.

The report was known, after its chairman, as the Aird Report. Its recommendations were delayed by the depression. The money to build and buy stations for a national broadcasting system was simply not available.

THE CANADIAN BROADCAST COMMISSION

In 1933, the government of Canada acted and created the Canadian Radio Broadcasting Commission (CRBC) which had a short life, but did get network broadcasting started in English and French.

The broadcasting Act of 1936 replaced the troubled CRBC with the Canadian Broadcasting Corporation (CBC), which still survives and thrives as Canada's national broadcasting system. To begin with, CBC had two main responsibilities: (a) to provide a national radio service for all Canadians; (b) to regulate all broadcasting, private as well as public.

For the next two decades the private stations fought regulation by CBC, which they considered a competitor.

In 1958, a separate regulatory body was established, the Board of Broadcast Governors (BBG). It performed this function, with some difficulties, for 10 years.

CRTC ESTABLISHED

In 1968, another change in the Broadcasting Act replaced the BBG with the Canadian Radio-Television Commission (CRTC). This body still exists as the Canadian Radio-Television and Telecommunications Commission (still CRTC), with certain telephone and other telecommunications regulation added to its responsibilities.

The present body, the CRTC, has regulartory authority over private radio and TV stations and cable and interprovincial telephone activities, a somewhat more limited authority over the CBC, which has its own board of governors, and reports separately to parliament.

BROADCASTING DURING THE WAR

The CBC was barely getting started when World War II started. The Corporation rose to the challenge. It sent mobile units to Europe with correspondents who reported the activities of Canadian troops in French and English. During the war, the CBC national news won a loyal following and CBC became a respected national institution.

POSTWAR BROADCASTING

After the war, the CBC continued its two-pronged growth, extending its transmission facilities from coast to coast in English, and covering Quebec and areas outside Quebec where French is spoken. It also developed its programming strength, extending from news and public affairs to attain excellence in farm broadcasts, educational programs, drama and entertainment.

THE PUBLIC-PRIVATE LINK

There is an important link between the public and the private sectors. From the outset the CBC has relied on a considerable number of private stations which serve as *affiliates*, carrying many of the corporation's programs and providing coverage to areas which the CBC stations do not reach.

These private affiliates, in both radio and television, provide links between the public and private sectors which are important to both.

THE RADIO YEARS

CBC was a radio broadcaster solely from 1936 to the arrival of television in 1952. Today, CBC remains an important radio broadcaster. After 25 years and more of television, many Canadians maintain that CBC radio is still the best side of the corporation. In both national and local radio, in both AM and FM, CBC has many loyal followers of its news, public affairs, music, drama and magazine format programs.

TELEVISION

Television arrived in 1952. Following the recommendations of the 1951 Massey Report, to begin with the CBC had exclusive coverage of the main centers. But soon, private stations were being licensed outside the main centers, and by 1960 were competing with CBC in the main cities.

In 1961, CTV, a national private, cooperative English-language television network was authorized. Later, Global TV, an English television network covering the densely populated part of Ontario was approved. In Quebec, TVA, a small 3-station French-language network, was approved.

In 1965, color television was authorized, and a decade later, about 50% of households had color sets.

CABLE TELEVISION

Cable TV is now a major factor in Canadian broadcasting. It began in the early 1950s, bringing in U.S. television signals even before Canadian television started. But, the real development came in the 1960s and 1970s.

Today, Canada is, per household, the most cabled country in the world. It is served by over 400 companies each with an exclusive license to serve a specific area, principally urban.

Canadian cable TV companies make no payment for the signals they take off the air. They are licensed and regulated by the CRTC, which encourages them to engage in community programming. These companies, particularly in the larger centers where they number thousands or hundreds of thousands of subscribers, are generally highly profitable.

A MIXED SYSTEM

As it has been since the 1930s, Canadian broadcasting remains a mixed system, partly public, partly private. The CBC is mainly financed from the public purse. Private stations and networks are almost totally financed by advertising. An effort is made to expand the total system so that programs will be provided in English and French to the total population.

The paradox is this: Canada is one of the best equipped and served broadcasting countries in the world. But broadcasting continues to be a subject of great controversy. The struggle continues between national and provincial governments, between proponents of public and private broadcasting, demands of native French peoples and other minority-language groups, rival claims to the use of new technology such as satellites and pay television.

The importance of broadcasting to Canadians, as evidenced by the nonstop controversy, can only be considered a healthy sign.

THE COMPONENTS OF CANADIAN BROADCASTING

Canadian broadcasting is composed of the CBC services, public educational broadcasting, and private.

Public Broadcasting—General

Canadian Broadcasting Corporation (CBC)*
1500 Bronson Avenue
P.O. Box 8478
Ottawa, Canada KIG 3J5

Names and addresses are provided. Readers may obtain up-to-date information by writing for annual reports.

This is the national broadcasting system, established in 1936, and it provides basic public broadcasting in English and French, by radio and television.

Stations owned and operated by CBC are supplemented by private stations which join the national networks as affiliates for an important part of each broadcasting day.

AM radio service is by national network and reaches almost the entire population in English and French, 98% of households.

FM radio is also in English and French, but it is limited to 83% of households.

Television, combining CBC-owned-and-operated stations with private TV affiliates, reaches some 97% of households in the two official languages.

Also operated by the CBC are three special services:

● *Northern Service*, which includes broadcasts in Inuit (Eskimo) and Indian languages as well as in English and French for the far north.

● *Armed Forces Broadcasting*: In conjunction with the armed forces, broadcasts for Canadian troops stationed with NATO in Europe, and with various peacekeeping forces around the world.

● *Radio Canada International* is Canada's overseas broadcasting service using shortwave and recordings to carry the voice of Canada in 11 languages to North and South America, the Caribbean, Europe and Africa. (See Figs. 7-1 and 7-2.)

In all its operations, CBC employs some 12,000 persons. In 1977-78, it reported total expenditures of $550 million. Of this amount, $91.8 million was derived from advertising revenue; the balance from government funds.

Economic pressure exists due to the constant increase in engineering costs in order to improve coverage, and the costs of producing more and better Canadian programming particularly for television. Government economy campaigns have added further to the CBC's financial problems.

Educational Public Broadcasting

Educational broadcasting dates back to the beginning of broadcasting. For about four decades, the main agent was the CBC, working in conjunction with the provinces which are responsible for education under the Canadian Constitution.

During the 1970s, three provinces set up their own educational TV corporations and have largely taken over educational TV operations in Ontario, Quebec and Alberta.

The CBC continues to do a limited amount of national radio and TV educational broadcasting in conjunction with the Council of Ministers of Education. It also continues to cooperate in providing facilities and assistance to those remaining provinces that have not set up their own corporations for educational TV.

TV Ontario; Ontario Educational Communications Authority, 2186 Yonge Street, Toronto, Ontario, Canada, is the largest and most active educational broadcasting system, with its own stations in southern Ontario and extending to other parts of the province as finance permits. It serves all age groups, and presents 16% of its programs in French.

Radio Quebec, 100 rue Fullum, Montreal, Quebec, Canada H2K 3L7, Radio Quebec: Office de Radio Telediffusion de Quebec, is an all-French service. It aims at general information as well as school broadcasts, and broadcasts sessions of l'Assemblee Nationale, of Quebec. It operates transmitters in Quebec, Montreal and Hull.

ACCESS: Alberta Educational Communications Corporation, 16930-114 Avenue, Edmonton, Alberta, Canada T5M 3S2, rents time on private TV stations rather than operate its own transmitters. It engages its program production at two production centers.

Private Broadcasting

The Canadian Association of Broadcasters (CAB) is located at 165 Sparks Street, Ottawa, Canada. The private broadcasters have led their own associations since 1926, and the majority of private radio and television station operators maintain membership in the CAB and in a series of affiliated regional organizations.

The association is spokesman for the private operators at public hearings of the CRTC, of parliamentary committees and Royal commissions. It also arranges exchanges of programs between members and provides other services for them.

In 1977, some 419 private AM and FM stations produced air time sales of $269 million. Private TV stations, numbering under 100, produced air time sales of $310 million.

Networks

A number of private stations joined forces in the early 1960s to form their own cooperative network, The CTV Television Network Limited (CTV), 42 Charles Street East, Toronto, Ontario, Canada M4Y 1T5. It now represents 15 TV stations and four affiliates and provides extensive coast-to-coast coverage in English.

Les Tele-Diffuseurs Associes (TVA), 1600 Maisonneuve Blvd. East, Montreal, Quebec, Canada H2L 4P2, links French TV stations at Montreal, Quebec, Chicoutimi, Sherbrooke, Trois-Rivieres, Rimouski/Sept-Iles, Rivere-du-Loup, Hull-Ottawa.

Global Communications Limited (Global), Global Television Network, 81 Barber Greene, Toronto, Ontario, Canada M3C 2A3, covers the densely populated portion of Southern Ontario with its own transmitters.

With minor exceptions, most radio network broadcasting is by CBC. In 1978, a company was licensed to provide an all news radio service. It is Canada All-News Radio Ltd. (CKO), 65 Adelaide Street East, Toronto, Ontario Canada M5C 1K6. By 1979, CKO had transmitters extending from Montreal to Vancouver, with plans to extend the network coast to coast as finance permits.

Fig. 7-1. Three 250 kW transmitters located at Radio Canada International's transmitter site at Sackville, New Brunswick, Canada.

Radiomutuel Network, Mutual Broadcasting Ltd., 1700 Berri Street, Montreal, Quebec, Canada H2L 4E8. This network joins French radio stations at Montreal, Quebec, Ottawa-Hull, Sherbrooke and Trois-Rivieres.

The Canadian Cable Television Association, 85 Albert Street, Ottawa, Canada, is the spokesman organization for the most of the more than 400 cable television companies now in operation. Its role has become more significant as their number and income has increased. A part of its current effort is to achieve pay TV for its members.

Licensing and regulation of the cable-TV companies, each of which has exclusive coverage in its assigned area, is the responsibility of the CRTC. In 1977, there were recorded a total of 427 operating systems, with direct and indirect subscribers totalling 3.4 million.

Regulation of Public Bodies

The Canadian Radio Television and Telecommunications Commission (CRTC), Central Building, Promenade du Portage, Hull, Quebec, Canada K1A 0N2, has an executive committee consisting of the chairman, two vice-chairman and six full-time members. In addition, there are 10 part-time members appointed on a largely geographic basis. The CRTC has a staff of about 400 to help carry out its duties of supervising and regulating radio, television, cable and telecommunications.

125

In several provinces, the telephone system is operated by the province or regulated by it. In the case of the largest system in the country, Bell Canada, which serves Ontario and Quebec, as well as in the case of B.C. telephones, regulation is provided by the CRTC.

CRTC took over from the B.B.G. in 1968 and a new broadcasting act gave it greater powers. In the area of policy, the CRTC has aimed at securing Canadian content in both radio and television, and in ensuring that ownership of the broadcast media is largely in the hands of Canadians. In the cable TV field, CRTC has encouraged operation of community programming channel by each cable system.

The ability of the CRTC to develop new policies to meet changing circumstances has not gone unchallenged. An important decision on Telecast, the satellite corporation was overturned by the privy council. New legislation for broadcasting has been proposed which would put important power in policy matters in the hands of the minister of communications.

CRTC was established in 1968 as a largely autonomous, semi-judicial body for determining matters in broadcasting, and since 1972, in telecommunication matters. At this point it is not possible to predict what changes, if any, will take place in the role of the CRTC.

Regulation of Private Bodies

The Department of Communications (DOC), Journal Tower North Building, 300 Slator Street, Ottawa, Canada K1A 0C8, is responsible for the technical approval for broadcasting license. Upon DOC approval, the grant is forwarded to CRTC, which is responsible for issuing the licenses. The DOC maintains an interest in the whole field of communications and encourages the growth of the Canadian electronics manufacturing industry. The CRTC reports to parliament through the minister of communications.

Regulation of Public Bodies

The Canadian Broadcasting Corporation (CBC), while highly autonomous, reports to parliament through the secretary of state, whose office is located at 15 Eddy Street, Hull, Quebec, Canada K1A 0M5.

The secretary of state is concerned with Canadian culture, involving responsibilities in the areas of bilingualism and biculturalism, as well as multiculturalism. These matters are related to broadcasting, as well as the arts, film and publications.

THE CURRENT SCENE

Canada is a communications and transportation country. With its vast land area, second only to the Soviet Union, and its slim, scattered population of 24 million, Canada is and has always been highly dependent on the means to move people, words and pictures.

In radio, television and cable TV, Canada is one of the best equipped countries in the world. Radio reaches about 98% of the population; television reaches 97% and in each case efforts continue to close the slim

Fig. 7-2. Radio Canada International's shortwave antennas located at Sackville, New Brunswick.

gap and make these media available to all Canadians. More than 400 cable TV companies reach 50% of Canadian homes, a world record.

From the 1930s, Canada has had a mixed system, partly publicily owned and partly private. The goal has been a unified system, so that the components complement each other to provide the best possible service to all. But the history of the past 50 years has been full of clash and conflict, and the battles are by no means over. Broadcasting was, is, and will be a subject of concern to Canadians. They have a number of divergent views on the best policies and practices.

Jurisdiction

The first enquiry into broadcasting, the Aird Report of 1929, raised the question of the participation of the provinces as well as the federal government in broadcasting.

The courts decided long ago that broadcasting was a federal responsibility. Education is definitely a provincial responsibility; therefore, a compromise has been marked out for educational broadcasting. But some provinces claim that cable TV takes place entirely within provincial territory and should be under their control. Other provinces have laid claim to all aspects of broadcasting because it is so vital to their cultural life.

Language

The national broadcasting system has responsibility for broadcasting in French and English. Private stations also broadcast in the two official languages.

While about one-third of Canadians speak French, they are not the only minority group with claims on the airwaves. The native people of Canada, the Indians and the Inuit (Eskimos), seek airtime to broadcast to their people in their own language.

Canada has welcomed immigrants from every part of the world and under the mosaic principle has invited them to preserve their ancestral languages and customs. There are several licensed radio stations and one television station in major cities which are designated multilingual and which broadcast in a multitude of languages including Italian, German, Ukrainian, Portuguese, Dutch, Japanese, etc.

Ownership

There is a division between public ownership and private ownership. National radio and television coverage is provided by the CBC, but with the aid of private affiliates.

While the emphasis of the private station is on local coverage, there are important exceptions. The CTV network offers close to complete national coverage in English television, based on a group of cooperating private stations. The Global TV Network, with its own transmitters, covers densely populated southern Ontario. The TVA joins eight private stations in Quebec.

Until recently, national radio was dominated entirely by the CBC. In 1978, an all-news English private radio network was established as CKO, Canada All-News Radio Ltd. At the time of writing, it extends from Montreal to Vancouver and is expanding to provide coast-to-coast coverage.

All cable TV is in the hands of private companies licensed by the regulatory agency, CRTC. In large centers, for a $5-$7 monthly subscription fee, the subscriber may receive up to 20 channels. Now, cable TV companies seek the right to introduce as a supplementary service, pay TV, in which the client will pay for the individual program or film, or for a special pay TV channel.

Another bone of contention is that the CBC television continues to gain a portion of its income from advertisements, although CBC radio is now noncommercial.

Two diverse groups would like CBC to drop TV commercials. Those who want to see more Canadian programs on CBC, advocate dropping imported U.S. programs which attract and carry ads. Some supporters of the private stations would like CBC to get out of advertising so that its competition would be eliminated and ads would go to the private stations and networks.

Technology

As in other countries, radio operators thought their world would come to an end when television arrived. But Canadian radio adapted and today is a healthy, thriving part of the broadcast scene.

So, too, with cable television. Television networks and stations fear they will be destroyed by the fragmentation of the audience. Despite cries of catastrophe, most television operations continue to thrive.

Each new technological advance sends shivers through the broadcasting community. A battle to serve the home looms between telephone companies and cable TV licensees. The use to which satellites should be put is hotly contested. Decisions on pay TV have been deferred, but there is general agreement that it is inevitable.

Revenue

With the exception of its earnings from advertising, the funds for the CBC come from the federal public purse and now amount to more than half a billion Canadian dollars annually. To meet the demands for more and better Canadian programs, even more funds are needed. Instead, in 1978, in the face of the need for austerity, the government made a substantial cut in the CBC's budget. Educational broadcasting activities are largely financed by the provinces, particularly in Ontario, Quebec and Alberta.

All other broadcast activities are private, financed from advertising revenue in radio and TV, and householder subscriptions in cable TV. In 50 years, the basic issues have not changed greatly, as has been noted.

BROADCASTING IN THE 1980S

Broadcasting has always been controversial, an indication of its importance in the lives of Canadians. It will continue to be a volatile area in the 1980s. These are among the forces that will bring about changes, the exact outcome of which is entirely unpredictable at this time. But changes there will be.

Political Change

After 16 years of liberal government, in May, 1979, Pierre Elliot Trudeau was toppled by the minority conservative government of Joe Clark. A change of government, and of ministers responsible for broadcasting, always brings new viewpoints, new actions. The Clark Government lasted little over six months and was replaced by a new government formed by Trudeau.

Federal Government vs Provinces

Quebec, with its Parti Quebecois, is not the only province demanding greater independence of Ottawa. Alberta, with its oil wealth, in fact all the western provinces, are talking greater control of their own affairs, under a revised Canadian constitution. So far, numerous meetings have failed to reach agreement on a formula for a revised constitution.

Because cable TV takes place entirely within provincial territory, several provinces are demanding complete control of cable TV, as well as future developments arising from it, such as pay TV. Some provinces would go far beyond this, and would seek control of all broadcasting matters.

Interest Groups

These have been present since the start of broadcasting. There are ideological differences between the proponents of public and of private ownership. The Canadian system is well established as a mixed system, with a firmly established nationally-owned Crown Corporation, the CBC, and equally well entrenched private radio, television and cable systems.

The demand of French-speaking Canadians from outside Quebec continues strong to improve and extend the French-language services available to them.

Programs in the Inuit and Indian languages are considered only a start by the native peoples, who constantly seek improvement in the number and quality of programs.

There are now several multilingual radio stations in Canada, plus a newly established multilingual television station in Toronto. In addition, the community programming which many cable systems offer, under CRTC encouragement, also features programs by local ethnic groups, often in the language of the country of their origin.

Most important and most unpredictable will be the impact of the technology on Canadian communications in the 1980s. The videotape and videodisc will end the isolation of the far north, where television entertainment is concerned. Videotape is already serving this end, at times illegally.

Satellite service is already in use, principally by the CBC, via the Anik satellite. (See Fig. 7-3). But private entrepreneurs have their eyes on satellite for such purposes as pay TV.

Testing of fiber optics promises an amazing new low-cost system of distribution and may increase the rivalry of telephone and cable TV in the competition to provide new services to the householder.

Fig. 7-3. Launching of the Anik satellite.

The cable TV system seems ideally suited to combine with the computer to pour endless quantities of information into the individual household on request. Visions of "wired city" and "wired world" are being paraded before our eyes.

The references here are to innovations which are already known, and some of which are already in use. Still newer developments, which are only about to appear, will certainly disturb the surface of Canadian broadcasting in the decade of the '80s. As in the past, even at the price of dislocations, Canadians will wish to take advantage of these innovations to overcome the isolation created by Canada's vast distances and small population.

Chapter 8
Broadcasting in
the United States

Dr. Bob Wallace
Edinboro State College
Edinboro, PA

Any discussion of the American system of broadcasting must be presented within the context of its economic base. Most of the issues facing American broadcasting can ultimately be traced to the fact that the dominant form of broadcasting in the United States is one that is commercial. American radio and television stations are businesses that seek the highest return on investment. The profit motive is crucial to most decisions.

COMMERCIAL BROADCASTING

As business ventures operating within the framework of a free-enterprise economic system, competition is an integral part of a station's daily existence. Proponents of commercial broadcasting argue that competition is the source of progress and is responsible in great part for most of the advances that have occurred. In the struggle to attract a larger share of the audience, broadcasters seek to enhance their productions with elaborate settings, unique special effects, improved color, and light-weight camera/recording systems.

The commercial system is also defended as the one system that is most responsive to the wishes of the public. In American broadcasting, a program will continue on the air only as long as it attracts a relatively large audience. By maintaining programs that appeal to the majority and discontinuing most others, the broadcaster contends that he is giving the public what it wants.

The possibility of profits has served as an incentive to businessmen, motivating them to invest their capital with the hope that they can successfully compete against other, established stations. Survival has often demanded that a station's management find some unique means of attracting an audience, perhaps by providing a specialized format. The growth of American broadcasting has been tremendous. The United States has 732 commercial television stations and 7657 commercial radio stations. Nearly 98% of the American households have at least one television set, with almost 48% of these having more than one set. Color receivers are now in over 80% of the homes. (See Fig. 8-1.)

In comparison to the rest of the world, the United States has nearly a third of all radio transmitters and over 20% of all television transmitters. And yet, the United States has only about 5% of the total population of the world. For the American public, the large number of stations means program diversity. For example, television viewers in New York City can select from over 200 different programs on a typical day. Zanesville, Ohio, a much smaller market with only 28,000 homes with television, has access to a daily average of over 75 different programs.

Another common defense of advertising-supported broadcasting is that the programming provided is free to the viewer. Without the revenue from the sale of commercial time, the high costs of station operation and program production would have to be assumed by the public through some means. A tax-supported system of broadcasting could be subject to political abuse. Freedom of expression would be endangered should the power of the purse enable government to determine program content. In surveys measuring public opinion regarding broadcast advertising, most Americans, although critical of certain factors, have indicated that commercials are a fair exchange for free programs.

The use of American airwaves for the marketing of products is not without controversy. Critics of the commercial system contend that the purpose of commercial broadcasting is not to present programs to the public. Programs are only bait. The real goal is to attract an audience for the commercials. The larger the audience, the more the station can charge for its advertising. As a result of the relationship between audience size and advertising revenue, a station presents mostly those programs that have mass appeal. Minimal efforts are made to meet the needs of minority interests, especially in areas such as fine arts and public affairs. Thus, when new programs are created, they imitate the same elements that have made other programs popular. In the sense that commercial broadcasting's programming strategy revolves around providing the advertiser with the largest possible audience, the public is not really being given what it wants.

A more extreme form of criticism comes from those who view commercial broadcasting as a part of the American capitalistic power structure. Broadcasting is not only a business, it is a *BIG* business. The networks and some stations are a part of large corporations with other enterprises. The interaction between industry and broadcasting (and other mass media) has made America a consumer society. Broadcast advertising is persuasion that conditions the public into becoming perpetual buyers of goods by relating the satisfaction of basic psychological drives and emotional needs to the purchase of the product. Happiness, love, security, respect, prestige, health, and sex are obtainable through consumption. The possession of objects is a measure of one's worth and success. And, finally, critics of commercial broadcasting refute the claim that free programming is provided since advertising is an expense that must be factored into the price of the product.

Opinion regarding the value of advertiser-supported broadcasting is obviously divided. The rightfulness of any side of the issue is one that is difficult to determine. Those favoring a given point of view will tend to overstate their case. And, the merits of an argument cannot be judged without one's own bias coming into play. Most people are predisposed to accept the validity of those ideas that support the ones that have already been established. Consequently, the complete resolution of the argument is not likely.

REGULATION

Although broadcasting emerged as a free-enterprise business, radio and television stations are and have always been subject to federal regulation. The airwaves are considered public property and cannot be owned by an individual. Therefore, a station cannot go on the air unless permission to do has been given by the government. As a condition for being granted the priviledge of using the public's airwaves, broadcasters are held accountable for their actions. They are trustees of public property and are obligated to serve the public interest.

Government Regulation

At the risk of oversimplification, the regulation of broadcasting can, then, be understood as a matter of insuring that stations do serve the public interest. The responsibility for overseeing station performance rests with the federal government through the United States Congress. Authority for congressional control is rooted in the United States Constitution. According to Article 1, Section 8, Congress is empowered to regulate interstate commerce. Broadcasting is classified as interstate commerce. Currently,

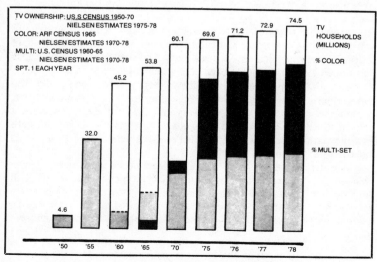

Fig. 8-1. Breakdown showing TV penetration of US households.

broadcasting is regulated under the provisions of the Communications Act of 1934. This Act of Congress created the Federal Communications Commission (FCC) as the agency to regulate stations on a day-to-day basis. The commission is comprised of seven members, each serving for a term of seven years. They are appointed by the President with the approval of the senate. No more than four members of the Commission may belong to the same political party. (See Fig. 8-2.)

The responsibilities of the FCC are quite broad. On the one hand, the commission is a legislative body by virtue of its authority to create the rules that govern broadcasting. These rules are not made without a great deal of deliberation and input from a wide number of sources. (See Fig. 8-3.) Notice is given by the FCC that it is considering the formulation of new or the revision of old rules. The opportunity for response and commentary from both the public and from the broadcasting industry is provided. Some specific rules will be discussed later in this section.

The FCC administers and enforces its regulations and it also acts as a judge in the event that a rule has been violated. However, its executive/judicial actions are tempered by the basic right of a station to "due process." A station is entitled to a hearing at which time testimony is presented. Actions by the commission can also be taken to the United States Court of Appeals. As a penalty for violating a rule, a station can be fined up to $10,000. Most minor violations result in either warnings or small fines. For more serious charges, a station may lose its license.

The power of the FCC to grant or revoke a license is its greatest source of control and authority. Licenses are issued for a period of three years. At license renewal time, the commission considers past performance, including complaints lodged by the public against a station and any possible rule violations. In some instances, the commission may decide to issue a short-term renewal as a compromise between either full renewal or license revocation.

In order to receive a license, an applicant must sign " . . . a waiver of any claim to the use of any particular frequency . . ." Broadcasters do, of course, own the physical assets of a station. And even though the license is not owned, it can be sold to another company—with FCC approval. However, the license must be held for two years before it can be sold. The application is made with the knowledge that the " . . . public interest, convenience, or necessity would be served . . ." In fact, the applicant must indicate that he has ascertained the needs of the area to be served and state how the station's programming will serve these needs. Fiscal soundness must also be established.

It is also the responsibility of the applicant to specify the frequency to be used. This means that the applicant must do an engineering survey to determine if the desired frequency is available without causing interference with other stations. A license cannot be issued to an alien, a foreign government or a corporation chartered under the laws of another country.

History of Regulation

The concept of public ownership of the airwaves and the subsequent need to apply the public-interest standard in issuing licenses is derived in part from the simple fact that frequency space is limited. Almost from the very beginning, the problem of scarcity plagued broadcasting. As more and more stations went on the air, interference among stations increased. The confusion included several stations broadcasting on the same frequency, power changes, and even frequency changes. Eventually the situation was so bad that governmental involvement was sought, something almost unheard of in free enterprise.

Some regulation of radio communications already existed. The Wireless Ship Act of 1910 mandated that a ship with 50 passengers must have a working radio communications apparatus before it could leave port. The Radio Act of 1912 came next. Although this act was an outgrowth of the Third International Radio Conference, the sinking of the Titanic gave emphasis to the need for regulation. Hundreds of lives were unnecessarily lost because the radio operators on nearby ships had signed off for the evening. This tragedy led to regulation requiring ship radios to be in continuous operation.

Over the next 15 years, radio broadcasting would undergo tremendous growth. "Ham" (amateur) radio operators increased in number. Experimentation began with point-to-point "wireless" radio transmission. Historians often credit Frank Conrad with establishing the first radio station in 1920. Conrad, a Westinghouse engineer involved with experimental radiotelephony, transmitted signals from his garage in Pittsburgh to an assistant at another location. Eventually, he included music in his broadcasts and developed a following that began requesting that certain songs be played.

Westinghouse was impressed with the public response and built a special transmitter. Westinghouse's action was motivated by its assumption that the public would need receivers in order to receive the station, receivers that Westinghouse would manufacture. The company thought that the profit in radio broadcasting lay in the scale of receivers. Stations would exist as a means of stimulating public interest and consequently the sale of receivers.

Conrad's experimental station, 8XK, became KDKA. Other stations were established at about the same time, including WHA in Madison, Wisconsin, and WWJ in Detroit, Michigan. WBZ in Springfield, Massachusetts, was the first station to be issued a regular license. Accordingly, KDKA's claim to being the first broadcast station is a matter of dispute. By 1925, over 570 stations were on the air, not including numerous homebuilt stations operating without authorization. Regulation was under the jurisdiction of the Department of Commerce and its secretary, Herbert Hoover. The shortage of spectrum space forced Mr. Hoover to issue licenses on the condition that an assigned frequency would be shared with another station. Court rulings eventually reversed this policy, declaring that any applicant meeting the requirements of the Act of 1912 must be granted a license and the authorization to engage in full-time broadcasting on an assigned frequency. Interference once again became a major problem.

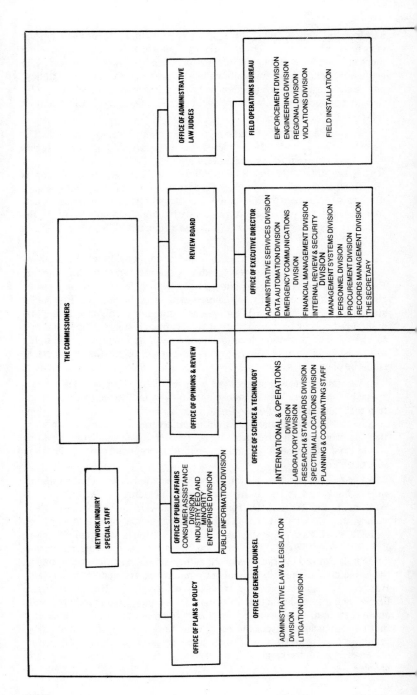

THE COMMISSIONERS

NETWORK INQUIRY SPECIAL STAFF

OFFICE OF PLANS & POLICY

OFFICE OF PUBLIC AFFAIRS
CONSUMER ASSISTANCE DIVISION
INDUSTRY EEO AND MINORITY ENTERPRISE DIVISION
PUBLIC INFORMATION DIVISION

OFFICE OF OPINIONS & REVIEW

REVIEW BOARD

OFFICE OF ADMINISTRATIVE LAW JUDGES

OFFICE OF GENERAL COUNSEL
ADMINISTRATIVE LAW & LEGISLATION DIVISION
LITIGATION DIVISION

OFFICE OF SCIENCE & TECHNOLOGY
INTERNATIONAL & OPERATIONS DIVISION
LABORATORY DIVISION
RESEARCH & STANDARDS DIVISION
SPECTRUM ALLOCATIONS DIVISION
PLANNING & COORDINATING STAFF

OFFICE OF EXECUTIVE DIRECTOR
ADMINISTRATIVE SERVICES DIVISION
DATA AUTOMATION DIVISION
EMERGENCY COMMUNICATIONS DIVISION
FINANCIAL MANAGEMENT DIVISION
INTERNAL REVIEW & SECURITY DIVISION
MANAGEMENT SYSTEMS DIVISION
PERSONNEL DIVISION
PROCUREMENT DIVISION
RECORDS MANAGEMENT DIVISION
THE SECRETARY

FIELD OPERATIONS BUREAU
ENFORCEMENT DIVISION
ENGINEERING DIVISION
REGIONAL DIVISION
VIOLATIONS DIVISION

FIELD INSTALLATION

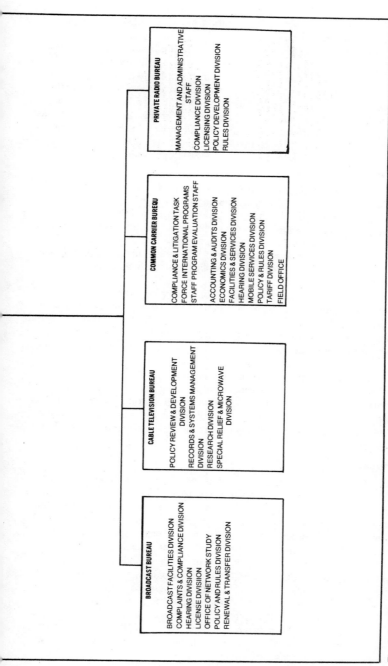

BROADCAST BUREAU

BROADCAST FACILITIES DIVISION
COMPLAINTS & COMPLIANCE DIVISION
HEARING DIVISION
LICENSE DIVISION
OFFICE OF NETWORK STUDY
POLICY AND RULES DIVISION
RENEWAL & TRANSFER DIVISION

CABLE TELEVISION BUREAU

POLICY REVIEW & DEVELOPMENT
DIVISION
RECORDS & SYSTEMS MANAGEMENT
DIVISION
RESEARCH DIVISION
SPECIAL RELIEF & MICROWAVE
DIVISION

COMMON CARRIER BUREAU

COMPLIANCE & LITIGATION TASK
FORCE INTERNATIONAL PROGRAMS
STAFF PROGRAM EVALUATION STAFF
ACCOUNTING & AUDITS DIVISION
ECONOMICS DIVISION
FACILITIES & SERVICES DIVISION
HEARING DIVISION
MOBILE SERVICES DIVISION
POLICY & RULES DIVISION
TARIFF DIVISION
FIELD OFFICE

PRIVATE RADIO BUREAU

MANAGEMENT AND ADMINISTRATIVE
STAFF
COMPLIANCE DIVISION
LICENSING DIVISION
POLICY DEVELOPMENT DIVISION
RULES DIVISION

Fig. 8-2. Federal Communications Commission Organization Chart (1979).

In an effort to improve the situation and to provide more comprehensive and coordinated regulation of all forms of radio communication, Congress passed the Radio Act of 1927. The Federal Radio Commission (FRC) was created as a part of this legislation. During its relatively short existence, the FRC concentrated on bringing order to the confusion on the air. However, effective regulation was not achieved. Wire communication, for example, was still under the authority of the Commerce Department. As a result of the inadequacies of the Radio Act, congress passed the Communications Act of 1934. All forms of electronic communication were now regulated by the same agency, the FCC.

Regulation of Content

Even though stations are required to serve the public interest, the FCC does not specify the programming that would meet this objective. Within the established concept of freedom of speech, it has long been held that governmental regulation of communications content—censorship—is inappropriate. Any such action would provide the government with the opportunity to control the free flow of information and possibly to shape the opinions and values of its citizens. Open communication is an essential part of American society.

Despite concern for the dangers inherent in governmental regulation, some segments of American society do not believe in unlimited self-regulation. A viewpoint exists that broadcasting is so pervasive that its impact on the opinions and thinking of the typical person is quite significant. On an average, the television set is on for over six hours per day in American homes. (See Fig. 8-4).

For many Americans, broadcast news is their major source of information. Without regulation, broadcasters could structure program content to serve their own self-interest. A station's presentation of controversial social and political issues could favor one side over another. Political candidates expounding the broadcaster's own ideology could get preferred treatment.

Fairness Doctrine

Several rules have been established by the FCC to insure that the public interest is protected when controversial issues are presented. Under the *fairness doctrine*, programming concerned with issues of public importance and controversy must be presented in a fair and balanced manner. Each program need not be completely balanced, but fairness must result in the long run. The opportunity must be available for differing points of view to be expressed. Under certain conditions, especially a personal attack on an individual, the broadcaster must actively seek out the other viewpoint. The fairness doctrine applies not only to station-produced programs, but also to those received from outside sources.

A station may use its own good judgment in selecting the person to be given air time to present the opposing viewpoint and need to make time available to everyone who requests it. In an instance of a personal attack, the person attacked has the right of reply.

Compliance with the fairness doctrine means extra effort and "aggravation" for the broadcaster. Differing opinions must be sought out and valuable time is lost when those viewpoints are aired. Not surprisingly, then, broadcasters are not inclined to make controversial programs an extensive part of their schedule. However, stations cannot completely

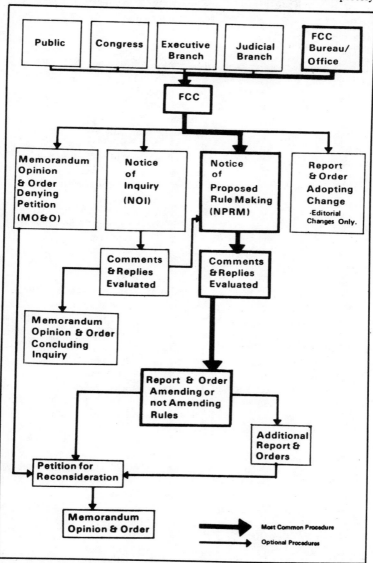

Fig. 8-3. Diagram explaining how FCC rules are made.

141

avoid such programming. As mentioned earlier, they are required to ascertain local problems, issues, and needs. Station executives *must* seek the opinion of community leaders about local problems. Then, programming to the issues must be proposed. The FCC evaluates the proposed programming whenever it considers an application for a license. In the event of a license renewal, the commission will compare what was promised with a station's actual performance.

Equal-Time Provision

Section 315 of the Communications Act of 1934 requires that should any ". . . legally qualified candidate for any public office . . ." be allowed to use a station, the licensee must provide "equal opportunities to all other such candidates for that office . . ." This regulation, often referred to as the equal-time provision, prevents a station from showing favoritism to a particular candidate and possibly influencing the results of an election.

Equal opportunity means providing both the same amount of time and the same quality of time at the same costs. Therefore, stations selling 100 commercial announcements to be broadcast on a highly viewed or heard program could not limit an opposing candidate for the same office to only 50 announcements to be presented at a less desirable time period.

The equal-time provision also applies to situations other than paid announcements. If a legally qualified candidate appears, without charge, as a guest on an interview program, then all other candidates must be given the same opportunity—if they so desire. This requirement is a particular problem for candidates who are employed as radio and television performers, e.g., actors and news reporters. Their on-air work is considered "use" of the station's facilities and, consequently, opposing candidates are entitled to equal time. A station will usually relieve an employee of any on-air job responsibilities until the election has been held. (Behind-the-scenes work, such as writing, is permitted). The performer-candidate could continue the on-air work if the station decided to give equal opportunity to the other candidates (not likely) or the opponents would agree to forfeit their claim to equal time (also unlikely).

The requirements of Section 315 apply to only those individuals who are legally qualified candidates. To be qualified, a formal announcement of candidacy must be made and all legal requirements for holding office (citizenship, residency, and age) must be met. News programs and news interviews are also exempt from the provision. However, a news program must be bona fide and shown on a regular schedule. A news editor can show as much of one candidate and as little of another as may be necessary in a given newscast. Obvious, blatant favoritism would most likely be subject to FCC inquiry and investigation.

A station's obligation to provide equal opportunity does not mean that every candidate will receive an equal amount of time. If the financial resources of one candidate are greater than those of his opponent, and he can buy more time than they can, the station is required to provide the other candidates with only the opportunity to buy equal time.

The impact of broadcasting on American political life is a matter of concern and is a source of debate regarding regulation. A well-designed

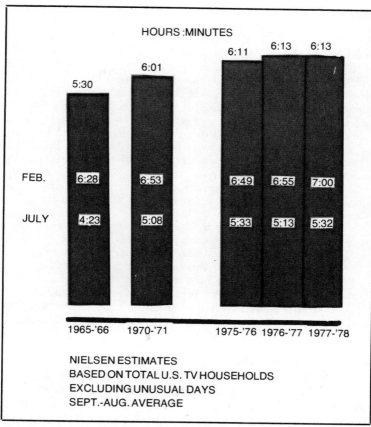

Fig. 8-4. Chart showing the daily TV viewing habits in the average US household (courtesy A. C. Nielsen Co.).

advertising campaign has become standard procedure for those seeking election to a public office. This is especially the case in national elections.

Political advertising is defended as an efficient means of presenting a candidate's position on election issues to a large audience. The result is a well-informed electorate, better prepared to engage in self-governance. News programs provide extensive coverage of election activities. During presidential elections, cameras follow the candidates around the country. Reporters give nightly updates of what each candidate said or did. The drama and excitment of the struggle to win unfolds before the viewer in his living room.

The extent that broadcasting adequately presents the issues is subject to discussion. A one-minute commercial can hardly begin to discuss the many ramifications of complex issues. News reports may pay too much attention to attention-getting, dynamic elements such as parades, crowds, and campaign rhetoric. Candidates attempt to manipulate the media by staging events or manufacturing news.

The substitution of image for content is another concern. Are candidates packaged and marketed like any other product that is sold to the public? A candidate, depending on the type of clothing worn (casual or a suit), the type of physical activity in which he is engaged (walking, sitting, or even riding a horse), and the location in which the commercial is filmed (farm, office, factory, street corner, woods, hospital, etc.) can be made to appear young, mature, calm, energetic, wise, sympathetic, pro-labor, pro-business, liberal, conservative . . . or any other image. A candidate may even be presented as many of these different images through the techniques of tailoring specific commercials to match the attitudes found in different geographic locations. The expectations of a candidate held by citizens in an eastern, industrialized city with strong unionism may be quite different from those held by farmers in a rural town in mid-America. (See Fig. 8-5.)

The issue of political advertising is a value judgment colored by one's own beliefs. Long-term exposure to both standard commercial advertising and to political rhetoric has made many Americans cynical toward any form of propaganda. Selective perception and pre-existing bias also play a role in how an individual reacts to a contrived image. In any event, it would certainly be a violation of freedom of expression for the government to ban political advertising or to specify how messages should be structured. Fair and equal treatment of all candidates is about the limit of governmental regulation.

Limits on Station Ownership

Since the scarcity of spectrum space limits the number of licenses that can be issued, those who are fortunate enough to operate stations are in a unique position of power and influence. The FCC recognized the potential peril that could result from the concentration of station ownership in the hands of a few individuals and established restrictions on the number of stations that could be licensed to one individual (or corporation). No more than seven television stations (five VHF and two UHF), seven AM stations, and seven FM stations may be licensed to the same owner. (Noncommercial broadcasting is exempt from these requirements.) Ownership within the same market is limited to no more than one station of the same kind. In order to have a greater exchange of ideas within a community, a broadcaster may not also own a newspaper or cable television in the same community to which his station is licensed.

Advertising and the Government

The FFC shares its regulatory authority with the Federal Trade Commission (FTC). Even the United States Congress investigates advertising practices. The FTC has dominant jurisdiction in situations involving fraud and deception. Should a commercial be found to be deceptive, the FTC can issue a cease-and-desist order requiring that the announcement no longer be broadcast. A business may also be fined and perhaps ordered to make restitution to consumers. However, misleading practices often are willingly stopped by an advertiser in order to avoid either legal fees or a fine. A station will not be penalized if it did not have any direct involvement in originating the deception. This is quite likely since many commercials, especially those for national and regional products, are created by advertising agencies, who in turn provide them to the station.

The FCC requires that stations identify the sponsorship of any material that is broadcast. In addition to standard, paid commercials, sponsorship also means the exchange of services. For example, the prizes given away on a game show are usually provided to the producer free of charge or at reduced cost in exchange for the on-air publicity received when the product is displayed and described. This type of "sponsorship" must be identified. The FCC is also concerned with the amount of time a station devotes to advertising. Although regulations do not specify the maximum allowed, the FCC has made it clear that it considers excessive advertising not to be in the public interest.

In recent years, the effect of advertising on the public's health has been a matter of concern and regulation. In 1972, after the surgeon general of the United States issued a report citing the potential health risks of cigarette smoking, cigarette advertisements were banned from broadcasting. (However, this restriction was not applied to the print media).

Debate regarding regulations on advertising directed to children has been on-going for a number of years. Lately, particular attention has been given to the possibility of placing restrictions on commercials for products with high sugar content, especially candy and sugar-coated cereals. These

Fig. 8-5. Scene of network coverage of national elections (courtesy ABC-TV).

products are heavily advertised on weekend cartoons. The major concern of those favoring regulation is that the commercials encourage poor diet, bad eating habits, and tooth decay.

On a broader scale, the total ban of *any* advertising aimed at children is supported by those who contend that young children are immature and not always able to make the best judgment. Advertising directed to them is a matter of exploitation. And, repeated exposure to advertising during a child's developmental period also conditions the child for a lifetime of conspicious consumption.

Broadcasters maintain that any regulation placed on them but not on other mass media is discriminatory and unfair, especially since stations normally lose advertising revenue to its print competitors as a result. The fact that broadcast programs are openly accessible to the public is a common justification for discrimination. Other mass media (books, magazines, film, theatre, and newspapers) cannot be received without a conscious effort, such as paying the price of admission. In the case of broadcasting, a young, impressionable child can turn on a television set without knowing the nature of the content about to be broadcast. Offensive material is an intrusion upon his innocence.

A broadcaster, like any other individual citizen, must comply with existing civil and criminal laws. Any owner violating any of these laws, including obscenity, libel, and lotteries, is subject ot prosecution. Since good moral character is one consideration for receiving a license, a criminal conviction could have serious implications at license renewal time.

Self-Regulation

The power of the government to regulate is limited by the constitution of the United States, which prohibits Congress from making any law "... abridging the freedom of speech, or of the press ..." This protection is reaffirmed for broadcasters in Section 326 of the Communications Act of 1934. In this provision, the FCC is denied the *"power of censorship"*. These safeguards restrain the government from any prior censorship of program content and gives broadcasting almost unlimited programming freedom.

The restrictions placed on governmental regulation of content places a great responsibility on the broadcast industry to serve the public interest and to engage in good practices. Although this may be similar to having a wolf guard the sheep, self-regulation is appropriate in terms of freedom of expression and is also consistent with the principles of capitalism and free enterprise.

At its most basic level, self-regulation is a matter of the broadcaster respecting the values of the community he serves. This is not an easy task since standards of "good taste" are continually shifting, usually in the direction of increased sophistication and permissiveness. Self-censorship based upon an artificial standard is further complicated by the fact that the United States is a country with a diverse population living in a large geographic area. Community standards of what is proper will vary among the different sections of the nation. For example, a station located in a religiously conservative community would be more sensitive to elements such as profanity and nudity than would a station in a metropolitan area.

Much of a station's programming originates from outside sources, usually a network. This presents a particular problem since local stations have no direct control over program content of network programs. A network will occasionally provide its affiliates with a preview of a potentially controversial program. If a station deems it to be unsuitable for its own community, it may choose not to broadcast the program. The right of a station to reject a network program is mandated in FCC regulations. A licensee must always have complete control over the programming that is presented.

Before a network presents any program, it has already gone through a process of self-censorship. Each network has a department that reviews programs, scripts, and announcements to insure that compliance is made with its established standards. As might be expected, the network "censor's" decisions do not always please the creative community. In the opinion of writers, performers, and directors, the network is too afraid of arousing public controversy and is, therefore, unwilling to take a chance on original story ideas and themes. The result is bland, antiseptic, and unrealistic programming. For the producer of a network program, production involves a regular routine of give and take with the network censor. It's a game of asking for more creative latitude than is really expected so that when a compromise is reached, the producer's original intent is achieved. It is also a matter of continually pushing against the established boundaries of restraint.

National Association of Broadcasters

Another source of self-regulation is the National Association of Broadcasters (NAB). The NAB is a professional organization in which membership is voluntary. Lobbying and professional development services are provided to its members. NAB has developed a code that specifies standards of good practice. The provisions of the code are somewhat general and relate to basic established values and standards of decency. The television code (a separate code exists for radio) has sections devoted to standards for both programs and advertising. According to the section, "Special Program Standards," a station should not exploit its presentation of violence, crime, and sex. (The presentation of explicit sexual acts is prohibited.) Care must be exercised so that those with physical handicaps are not ridiculed. Profanity and obscenity are prohibited.

The advertising standards prohibit commercials for liquor. The characters (real or animated) in a children's program shall not, on that same program, be used to sell a product. Health-related products shall not be advertised to children. In commercials for health-type products (over-the-counter drugs), professionals in the field of medicine and health services shall not be used. Standards are also specified for the amount of

147

time that can be devoted to "non-program material," including both commercials and promotional announcements. The maximum time permitted varies according to the time of day a program is broadcast and whether or not a station is affiliated with a network. In prime-time programs, non-program material is limited to 9 minutes, 30 seconds per hour.

Compliance with the code is voluntary. If the code is violated, the NAB can revoke the station's membership in the association. The decision whether or not to comply becomes a matter of not only conscience, but also economics. One requirement that a station must meet in order to be granted a license is fiscal soundness. However, not all stations are profitable. The struggle for survival may make adherence to a code economically impractical.

PROGRAMMING

Self-regulation is more than conforming to standards of decency and to matters of fairness in presenting controversial issues. A station must also regulate its total broadcast schedule so that a balance of programming is presented and the overall needs of the total audience are met. By implication, programming is a vital part of the licensing process. Given the fact of limited spectrum space and assuming competition for an available frequency, the selection of one applicant over another suggests a value judgment on the part of the commission. If every applicant can meet the fundamental requirements (fiscal soundness, citizenship, and moral character), then the determining factor becomes one of how the public interest will be served as reflected in the programming proposed by an applicant. American stations select their programming from among three primary sources: local production, network affiliation, and program distributors.

Local Production

Although each licensee promises to serve the public interest, the fact that commercial stations are businesses seeking a profit is a reality that demands primary consideration. Program decision-making is essentially a matter of income vs. expenses. Since the value of a station's advertising time is directly related to the size of its audience (the larger the audience, the more the station charges), programs are selected on the basis of their ability to attract the largest possible audience at the lowest possible cost. Production of quality television programs, such as drama and documentaries, is a high-cost enterprise. (A network prime-time situation comedy costs nearly $200,000 per episode.) The expense incurred in set design, script writing, professional performers, and technical requirements are beyond the range of what most local stations can afford to offer on a regular basis. Small-market stations, in particular, do not receive enough advertising revenue to support major production costs. (The size-of-audience/cost-of-advertising factor depends in part on the size of the community in which the station is located. Compare the costs for the network base hourly rate in the following communities: New York City, $10,000; Miami, $2027; Erie, $950.) Local stations usually limit their productions to newscasts, interviews, game shows and sports.

For radio, local production usually means a music-and-talk show featuring prerecorded music with a "live" announcer. Even this tradition is changing as more and more stations buy complete programs from outside sources and play them with automated switching equipment. Sports programming has long been a strong part of local radio. For the price of an announcer and a telephone call (long distance at away games) a station can capture a specialized and loyal audience. Another radio format, usually found in urban areas, is the all-talk station, featuring lots of news, interviews, and telephone call-ins.

Network Affiliation

Most stations, primarily television, have found it convenient to acquire programs by affilating with a network. Simply defined, a network is the interconnecting of several stations for the purpose of transmitting the same program at the same time. The basic working arrangement between a station and its network is simple. The station makes certain blocks of its time available to the network and receives free programs in exchange. Stations must give their clearance for each program since, as mentioned earlier, a licensee cannot transfer its programming authority. The network assumes all costs and service charges for the programs. Not only are the programs free, the station is compensated for carrying them. The compensation, only about 30% of the station's established rate, is paid from revenues the network receives from selling the time from its affiliates to national advertisers. The local station also receives another benefit. It can sell its own advertising time for commercials to be broadcast during "station breaks," the time interval between the end of one network program and the beginning of another when the local station announces its call letters. Network sponsorship offers the advertiser an efficient means of reaching national audiences, eliminating the paperwork involved in negotiating individually with each station. (See Fig. 8-6.)

Networks

Networks began, of course, in the early days of radio. Credit for the first network is given to the American Telephone and Telegraph Company (AT&T) who, on January 4, 1923, provided a 2-station hookup between WEAF (New York) and WNAB (Boston). In slightly more than a year, this network grew to 22 stations. A second network was soon begun by the Radio Corporation of America (RCA) with the interconnecting of WJZ (New York) and WGY (Schenectady). In 1926, AT&T decided to remove itself from the group of companies (AT&T, General Electric, and Westinghouse) that had founded RCA. Its network was sold to the National Broadcasting Company, a new subsidiary of RCA. RCA/NBC now owned two networks. Eventually, one network was known as the Red Network and the other as the Blue Network.

Another network, United Independent Broadcasters, was developing about the same time as NBC. Faced with financial problems, it sought additional funding from the Columbia Phonograph Broadcasting System. Financial problems continued, the phonograph company dropped out, and the network became the Columbia Broadcasting System (CBS). In 1928, the economic survival of CBS began when William Paley invested in the network and assumed majority ownership.

The Mutual Broadcasting System (MBS) was formed in 1934. The inroads and progress made by NBC and CBS made the growth of MBS difficult. For example, the other networks had been able to affiliate with the nation's largest and most powerful stations. Affiliation agreements at that time gave networks greater authority over station programming. The FCC decided that the power of NBC and CBS was so great that competition was restrained and issued the "chain broadcasting" regulations. According to one regulation, the FCC would not license a station affiliating with an organization having more than one network. As a result, NBC sold its Blue Network in 1943. The Blue Network became the American Broadcasting Company (ABC). During the 1940s and early 1950s, the Dumont network (owned in part by Paramount) provided a fourth network service. However, it was unable to survive the competition.

Unlike their station affiliates, networks are not regulated by the FCC. No license is needed to operate. However, a form of secondary regulation exists. The FCC specifies the conditions under which a station can enter into an affiliation agreement. The rules state that a license will not be granted to a station if its network agreement contains certain restrictions. For example, a station must maintain control over programming and have the right to reject network programs. A station cannot give the network the right to set the advertising rates for non-network programs. Stations cannot be penalized for carrying programs from other networks. A station cannot enter into an affiliation agreement for a period longer than two years. However, the agreement can be renewed.

All three networks are not only program suppliers, but also owners of stations licensed by the FCC. The network owned-and-operated stations are subject to the same FCC regulations that apply to all other stations. Since the character of the licensee is a factor in renewal, the networks are sensitive to the need to avoid gross violations of decency and fairness in their natural program service.

The control that networks have had over the programming of their affiliates has concerned the FCC for a number of years. In the early days of radio, advertisers exercised a great deal of influence on programming. But as television developed, the networks gained greater authority. Control was exercised in the form of partial ownership of some of the programs the networks placed on their schedules. Thus, when a program was eventually taken off a network and made available to local stations on an individual basis, the networks were in a position to influence a station's non-network program selection. The networks even established their own companies for the purpose of program syndication.

For the FCC, the concentration of programming in the hands of a few

Fig. 8-6. Views inside and outside ABC's mobile Unit 6, a $3 million investment that was built by ABC (American Broadcasting Co.) primarily for sporting events (courtesy ABC-TV).

sources was too great and it enacted the *prime-time access* rule in 1970. Networks were no longer permitted to have United States syndication rights to programs provided by independent producers. In addition, stations in the top 50 markets were not permitted to carry more than three hours of programming during prime time (7-11 p.m., Eastern Standard Time). Prior to this ruling, local stations usually received network programs from 7:30 p.m. until 11:00 p.m. (at 11:00 p.m., local news is presented for 30 minutes and then it's back to the network for nearly two more hours). The loss of the top 50 markets made it economically impractical for the networks to provide a program service to its other affiliates.

In effect, the prime-time rule applied to all stations, regardless of market size. By limiting the scope of network programming, the FCC intended, among other things, to generate an opportunity for independent producers to have a market for their programs. And, some observers thought that the extra time provided to the local station could be used to create local programs featuring local talent and discussions of the issues of each community.

Program Syndication

Network programs are presented in time blocks of two to three hours, with a one to two hour break between blocks. It is the responsibility of the local station to fill the "hole" between network presentations. A station could, of course, present local productions. As noted earlier, this is not a likely alternative. The usual procedure is to rent *syndicated* programs from outside distributors. The station is given exclusive rights to broadcast the programs. A station will often lease the rights to a series that was formerly on one of the networks.

Game shows are another popular format. After an episode has been presented, it is returned by mail and another station in a different community will then broadcast it.

Syndicated programs are received without commercials (with the exception mentioned below) and the local station must find sponsors if expenses are to be met and a profit made. Advertising time is sold both to local businesses and also to national advertisers seeking greater coverage than that provided by their network advertising. In the case of a local advertiser, the station will probably produce its own commercials. National "spots" are provided by advertising agencies. Advertising agencies provide several specialized services to their clients, including concept development, production supervision, campaign planning, and time/space buying. Agencies traditionally work on a commission basis.

Some syndicated programs are provided free to the station by the advertiser. These programs contain a certain number of commercials for the sponsor's product that must be broadcast without charge. There are also blank commercial time segments in which the station can run its own commercials. This procedure is known as the barter system.

Programming Strategy

Since the prime objective of a station or network is to reach the largest possible audience, how does a programmer go about meeting this objective? The answer is a combination of common sense and a form of gamesmanship called counter programming. It's a matter of common sense to schedule a program according to the type of audience that is available at a given moment of the broadcast day. For example, it would be a wasted effort to schedule programs intended for school-age children during the time of day they are in school. In addition, the early-morning routine of eating breakfast and rushing off to school or work requires that television programs be easy-going and not demand a lot of attention and involvement on the part of the audience. The networks, accordingly, begin their day with news and information programs that are divided into self-contained segments.

After the workers and the children have left the house, programmers must now exercise a little more critical thinking. Several realities must be recognized. Household chores compete with television for attention. The audience is relatively small, mostly housewives, young children and retirees. A small audience limits the expense that can be allowed for a production. Consequently, inexpensive programs such as game shows, romantic melodramas (known as "soap operas" since they are often sponsored by manufacterers of soap products), are presented.

At this point, specific programming strategies come into play. What does CBS schedule at 11:00 a.m. against NBC game shows and an ABC series repeat? The choice is probably between another game show or a repeat series, since it's a little too early in the day for a "soap opera."

"Soaps" have a very special place in the hearts of regular viewers. Some audience members develop a strong attachment to their favorite programs, almost to the point of becoming personally involved with the romantic triangles and miscellaneous tragedies that happen to the characters. Each daily episode usually ends in a "teaser" or "cliffhanger" and the viewer is forced to tune in again the next day in order to find out what is going to happen. Because of the strong viewer involvement, most soaps are scheduled for the afternoon when the housewife has completed many of her tasks.

The network's afternoon programming ends about the same time the children and workers return home. The local station is now faced with the challenge of providing its own competitive programming for nearly two hours. Should a station select children as its target audience? If so, what type of children's programs should be selected? Half-hour situation comedies (sitcoms) are popular and appeal not only to children but also to other family members. However, what should be done if another station is already broadcasting a sitcom? Is there enough audience for one more program of the same type? Perhaps a cartoon series would do better. Or, a station may decide to present an action-adventure series in order to attract viewers who have no interest in comedy programs. (See Fig. 8-7.)

The local program schedule, consisting mostly of syndicated material, will run until the early newsblock, a combination of 30 minutes of local news and 30 minutes of national news presented by the network. Then it's back to locally scheduled programs, usually more syndicated programs. Network programming resumes at 8:00 p.m. (Eastern Standard Time) and continues until 11:00 p.m.

The evening block is the time of day that television set usage is the greatest and is known as prime time. Radio prime time falls in the period known as *drive time*, the time of day when workers and older students are listening to their car radios while they are driving to or from work or school.

Most network programmers believe that it is very important to start the prime-time period with a strong attractive program (known as a *lead-in*) that will immediately capture a large audience. Hopefully, this same audience will stay tuned to the same station for the next program. Various techniques are used to discourage the audience from switching channels. A very popular one-hour program will be scheduled against two half-hour programs presented by another network. A third network may schedule a 2-hour movie against the one-hour program!

The selection of a network's lead-in involves the same type of thinking done by the local station when it schedules syndicated programs. Would a situation comedy do better than an action-adventure series? How well would a variety show do? One programming decision is certain: A network will hardly ever lead with a news-public affairs special. Since these programs usually have small audiences, they are scheduled at 10:00 p.m.

The first hour of prime time is known as the family hour, a time period when programs should be appropriate for all family members. In other words, not too much sex and violence. The family hour is a self-imposed restriction and is a part of the NAB Code. However, pressure to implement the idea came from the FCC. Compliance with the family hour eliminates certain types of programs from consideration as lead-ins.

At the end of prime time, local stations present the late evening news. Network programming returns 30 minutes later and will last until approximately 2:00 a.m. The networks entries at that time of night include a talk show featuring interviews with entertainment personalities, reruns of programs previously seen in prime time, movies, and occasional specials.

The type of program scheduling described above is typical of most week days and of the evening programming on weekends. Daytime programming on weekends consists of cartoons in the morning and sports events in the afternoon.

When, then, is the public interest served? Of course, there are the daily news programs. Stations in larger cities produce their own magazine-type programs in which interesting people and events are presented. Most public affairs programs, however, are broadcast during the day on weekends, especially on Sunday, where they can do little damage to a station's struggle to attract a large audience. Although this may be questionable within the public interest concept, the reality of the situation is that most Americans have little interest in public affairs programming, even well-produced network documentaries.

Radio Programming

At one point in its development, radio offered network programming as varied as that found on today's television: variety shows, drama, soap operas, comedy, game shows and sports. In fact, many early television programs originated on radio.

Once television became the dominant medium, radio had to adjust in order to survive. The public would rather watch a drama than listen to one. Most stations developed a specialized *sound*—music format designed to attract a specific audience. One station may play only "rock and roll." Another will concentrate on "country and western," and another on "easy listening."

In larger cities, more stations specialize. The larger population provides an audience base large enough to support stations offering classical, ethnic, or talk programming. There are even sub-sets of general program types. For example, rock music appealing to young children is known as "bubblegum" rock. Older youth may enjoy "progressive" rock.

In order to increase the probability of reaching its target audience and gathering its share of audience, a station may employ the services of a programming consultant. After analyzing the market and the competition, the consultant will make recommendations regarding format. The format

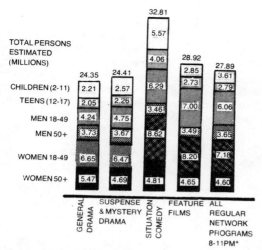

TOTAL PERSONS ESTIMATED (MILLIONS)

	24.35	24.41	32.81	28.92	27.89
			5.57		
			4.06	2.85	3.61
CHILDREN (2-11)	2.21	2.57	6.29	2.73	2.79
TEENS (12-17)	2.05	2.26	3.46	7.00	6.06
MEN 18-49	4.24	4.75			
MEN 50+	3.73	3.67	8.62	3.49	3.65
WOMEN 18-49	6.65	6.47		8.20	7.18
WOMEN 50+	5.47	4.69	4.81	4.65	4.60

GENERAL DRAMA · SUSPENSE & MYSTERY DRAMA · SITUATION COMEDY · FEATURE FILMS · ALL REGULAR NETWORK PROGRAMS 8-11PM*

*7-11PM SUNDAY EASTERN TIME

Fig. 8-7. Average audience composition of selected prime-time program types (Coutresy A. C. Nielsen Co.).

can be quite structured, requiring the program host (disc jockey) to play a certain type of music in a certain order according to the mix of sound specified by the expert. The formula takes into consideration the proper placement and contrast of differing tempos, vocal, and instrumental selections.

Another trend in radio, mentioned earlier, is the automation of programming. The entire musical programming of a station can be purchased on tape and may even include a prerecorded announcer. Should a station decide to use a live announcer, the program supplier can provide the station with the exact copy (script) that should be said. Provision is made for the station to insert local materials, including commercials, station identification, weather and news.

RATINGS

The success of a program is directly related to the size of the audience it attracts, its *ratings*. Research companies continually gather data on audience viewing and listening behavior. As a management information tool, audience measurement plays a vital role in determining the selection and scheduling of programs, the value of a station's commercial time, and the choice of time slots an advertiser will buy. Billions of dollars are at stake in the process of buying and selling time in the production of programs.

The significance of a program's rating is a matter of data interpretation. For example, the size of an audience can be measured in terms of *head count*, the total amount of people tuned in. Gross numbers, although better than nothing, have limited application. Additional meaning is provided by comparing the head count to the audience of other programs. Any comparison shoud be made within the framework of both the *potential* audience and the *actual* audience that was available.

155

The potential audience, in the case of television, refers to the number of homes in the area surveyed having receiving sets. Assume, for example, that a market has 100,000 households with television receivers. Potentially, all 100,000 households could tune in the same program at the same time. This is most unlikely. The audience that a program did attract could be compared to its potential audience by dividing the potential into the actual audience. The numerical result is the program's *rating*. For the sake of illustration, assume that a program had 30,000 viewers:

$$\frac{\text{ACTUAL}}{\text{POTENTIAL}} = \frac{30,000}{100,000} = .30 \text{ or } 30\%$$

Its rating would be 30. (Percentage symbols and decimal points are not used when expressing a rating.) Although stations are located in specific cities, their signals travel within a wide geographic area and often cover more than one city. Therefore, the primary coverage area of a station is considered within the "market" concept, rather than a city. The market area of a station, based on its signal coverage, is known as its *area of dominant influence* or ADI.

It is also unlikely that the television sets in all 100,000 homes were in use at the same time. The number of homes with their sets turned on, referred to as *homes using television* (HUT), represents the actual audience available to a program. The HUT percentage is calculated by dividing the potential audience by the number of sets in use. If, in the mythical market of 100,000 sets, only 60,000 sets were in use, the HUT would be 60.

$$\frac{60,000}{100,000} = 60\%$$

The HUT percentage is affected by variables such as time of day and day of week. As stated above, set usage is greater in prime time. Television viewing is also greater in winter. (Bad weather keeps people indoors.) If it is to be meaningful, any interpretation of a program's rating should be made within the context of the HUT concept. By dividing a program's audience by the number of sets in use, a comparison can be made of a program's ability to compete against other programs in the same time period. Its *share* of the audience can be determined.

Continuing with the hypothetical example, the share for a program having 30,000 viewers when 60,000 sets were in use would be 50:

$$\frac{30,000}{60,000} = 50\%$$

A share of 50 is dramatic when compared to the rating of 30. It means that the program was viewed by half of the available audience—a tremendous rating by most standards. Generally speaking, a program needs to capture at least a third of the available audience if it is to continue on the broadcast schedule.

Data Collection

Audience measurement is a process of survey research using a small sample of the total population. The sample is selected at random and, using United States Census data, is balanced (theoretically) to represent the total "universe." Particular attention is paid to representing demographic factors such as age, sex, education, income, occupation, and geographic location. Conclusions regarding a program's audience are mathematical approximations based on the laws of probability and statistics.

The actual procedure for gathering data is a critical factor in the reliability and validity of audience research. One common method of collecting data is the *diary*. Each home surveyed keeps a weekly record of every program selected, noting the time of day, station call letters and channel number, and program title. Demographic information relating to those watching (in the case of television) the program is also indicated. At the end of the week, the diary is mailed to the research company for tabulation. A major user of the diary is Arbitron (American Research Bureau). The A. C. Nielsen Company used a Recordimeter in conjunction with its diary. This electronic device records set usage so the verification of diary data can be made. It also emits a signal every 30 minutes (when the set is on) as a reminder to make diary entries.

The Nielsen Company also uses the Audimeter, an electronic meter attached to the home television set that continually records when the set is on, the channel selected and even the switching of channels. The Audimeter is linked by telephone to a central computer system. The computer "calls" each home Audimeter at least twice a day and retrieves the information. (See Fig. 8-8.) This process is instantaneous and does not involve any member of the participating household. (See Fig. 8-9.)

Audiences are also measured by *telephone interview*, especially in surveys of radio audiences. In the telephone *recall* method, a person is asked to remember what programs and stations were selected. The telephone *coincidental* procedure involves conducting the interview at the same time a program is being broadcast. This technique avoids the possibility of memory failure that can occur with the recall method. Telephone interviews are a relatively quick and inexpensive means of collecting data.

Criticism of Sampling Techniques

The validity of the diary method depends on the complete cooperation of those surveyed to continually and accurately record every program selection. Interruptions around the house and boredom with the procedure can occur.

If entries are made at a time period later than the actual program selection, will recall be accurate? Although a meter can measure when a set is on, it cannot provide demographic information. Nor can it detect those instances when a set is turned on and nobody is watching it. The telephone recall method is dependent on the interviewee's memory. Coincidental calls are an interruption of a person's activities and the person called may not be cooperative. The major concern that applies to all of the techniques used is *sample error*. Was a large sample used (Nielsen uses nearly 1,170 households with Audimeters) and was the sample truly representative?

Use of Ratings

Ratings provide advertisers with a means of determining cost effectiveness. Efficiency is often measured by dividing the cost of advertising time by the size of a program's audience. The result is expressed in terms of how much an advertisement costs per thousand homes (CPM). As a general rule, an advertiser seeks the lowest possible cost per thousand. Similarly, programs with a high CPM are difficult for a station to sell. If a program cannot attract a sponsor or if its time must be discounted (beyond standard volume and frequency discounts), it will probably be removed from the broadcast schedule.

Advertising strategy involves more than just the CPM factor. Ratings provide demographic information that must also be considered, especially if an advertiser is trying to reach a specific audience. For example, a dress manufacturer would need to know what programs were most watched by women. The target audience can be even more specific. A car manufacturer wants to advertise to men and women who are legally old enough to drive a car and who earn enough money to make car payments. Should the car be an expensive model, the target audience would be restricted also to those in the upper income levels. Sponsorship of a program with a slightly higher CPM but a greater concentration of the target audience may be more efficient than buying time on a program with a lower CPM but a more diverse audience composition. (See Fig. 8-10.)

General Criticism of Ratings

Ratings have been and will continue to be a source of controversy. Critics degrade ratings when a quality program is canceled because of insufficient audience. Directors, writers, and all other members of the creative side of broadcasting are quite vocal when ratings turn their work into failure. Instantaneous ratings provided by meters are blamed for contributing to a premature demise of programs that may have been successful if they had been given enough time to develop audience awareness. The concept of failure of a broadcast program is interesting when compared to the criteria for success for other mass media. A prime-time television program with eight million viewers would be considered a failure. A book or record selling eight million copies would be a great success.

Ratings are also accused of contributing to the lack of minority or special-interest programming. The target audience sought by most advertisers are those people in the mainstream of consumption. Typically, these are young and middle-aged adults who must buy food, clothing, shelter, and many other needs for their growing families. In contrast, the very young have neither money nor the means to spend it if they had any. The elderly, living on limited savings, pensions, or social security, have already gone through a lifetime of consumption and have entered a period

Fig. 8-8. Audimeter used by the A. C. Nielsen Co (Courtesy A. C. Nielsen Co.).

of relative purchasing stability. As a result, programs are selected on the basis of meeting advertiser needs. It should be no surprise that programs centering around the needs and interests of older citizens do not receive high priority.

Despite the questions that have been raised, the high costs involved in the business of broadcasting make ratings a necessity. The dollars involved in broadcast advertising are enormous—over 11 billion a year in total revenues. This is more than the GNP of some countries! Since not all programs will succeed in reaching a large audience, broadcast advertising is a calculated risk. Ratings provide one means of reducing that risk.

NONCOMMERCIAL BROADCASTING

Radio and television in the United States is not all situation comedies, game shows, and commercial advertising. A noncommercial system of broadcasting also exists. The beginnings of this system go back to the beginning of broadcasting in this country.

Radio

The University of Wisconsin was experimenting with radio in 1919. In 1923, educational institutions operated 72 stations. During the years between 1921 and 1936, over 200 institutions had received licenses. However, the number of educator-operated stations dropped to 38 by 1937.

A variety of factors led to the decrease of educational radio; among them:

159

program audience averages section

HOW TO READ THE PROGRAM AUDIENCES AVERAGES SECTION

The average quarter hour DMA HH ratings for *To Tell the Truth* on Mondays for the four weeks of the measurement period were 13%, 16%, 12% and 15% for weeks 1, 2, 3 and 4 respectively. The multi-week average rating was 14 which represents a 26 share. Note that the HUT level for Mondays 7:00 p.m.—7:30 p.m. was 54.

EASTVILLE

7.00P

R.S.E. THRESHOLDS) 25-% (1 S.E.) 4 WK AVG/50-%

METRO HH RTG	SHR	STATION DAY	PROGRAM	DMA RATINGS Wk1	Wk2	Wk3	Wk4	MW RTG	MW SHR	OCT 78	MAY 78	FEB 78	NOV 77	HUT	HH	W 18+	W 18-34	W 18-49	W 25-49	W 25-54	WKG	M 18+	M 18-34	M 18-49	M 25-49	M 25-54	CH 2-11	CH 6-11
12	22	WAAA MON	HOLLYWOOD CONN	11	11	13	11	11	21	20	27	20	20	54	72	60	14	29	27	35	15	36	11	21	17	21	14	9
10	19	TUE	HOLLYWOOD CONN	10	10	7	8	9	18	18	24	18	20	52	61	51	17	28	22	27	11	35	14	18	21	21	14	9
12	23	WED	HOLLYWOOD CONN	11	11	11	10	11	20	18	25	21	23	50	66	53	13	27	24	31	11	34	13	20	21	27	14	11
13	25	THU	HOLLYWOOD CONN	10	14	9	11	11	22	22	25	19	23	51	73	61	16	31	26	32	12	40	13	20	17	21	16	12
11	23	FRI	HOLLYWOOD CONN	12	12	12	10	10	20	19	27	21	27	49	62	54	12	28	24	29	14	36	12	17	14	19	13	10
11	22	*AVS*	*HOLLYWOOD CONN*	*11*	*11*	*10*		*10*	*20*	*20*	*26*	*21*	*22*	*48*	*67*	*55*	*14*	*28*	*24*	*37*		*38*	*12*	*22*	*17*	*22*	*14*	*12*
18	41	SAT	HEE HAW	16	22	18		20	38	40	38	35	39	62	124	100	24	41	34	51	24	92	24	44	35	49	37	21
<	24	SUN	LUNDY&STALLION	12				18	35	38	31	24	62	60	175	168	20	35	31	38	23	99	19	27	35	21	55	25
14	24	*NOR*	*WORLD-DISNEY*	*17*	*14*	*12*		*15*	*24*	*25*		*36*		*60*	*97*	*74*	*33*	*50*	*34*	*38*	*23*	*63*	*26*	*42*	*35*	*40*	*51*	*33*
14	24	*AVS*	*WORLD-DISNEY*	*18*	*14*	*12*		*15*	*24*	*29*			*38*	*61*	*95*	*72*	*30*	*46*	*40*	*67*	*22*	*65*	*25*	*34*		*40*	*51*	*33*
14	26	WBBB MON	CROSS WITS	14	14	15	14	14	26	19	24	26	22	54	89	75	17	29	23	30	16	55	12	20	19	25	13	8
14	26	TUE	CROSS WITS	15	15	15	12	12	26	21	25	24	20	52	89	75	17	31	22	29	15	49	11	19	17	21	14	7
13	25	WED	CROSS WITS	14	14	12	13	13	25	18	23	23	20	50	89	72	13	30	27	28	17	48	10	18	14	17	15	10
<		THU	CONV'S-MAYOR					13	23	10	29	24	23	64	64	72	9	17	9	13	12	39	12	11	4	11	11	6
13	26	FRI	CROSS WITS	15	12	13	15	13	26	20	27	27	21	49	82	78	24	35	27	34	18	54	14	22	19	25	11	8
13	26	*AVS*	*CROSS WITS*	*13*	*13*	*13*		*13*	*26*	*19*	*25*	*25*	*21*	*51*	*84*	*72*	*18*	*30*	*23*	*30*	*17*	*49*	*12*	*22*	*16*	*20*	*14*	*8*
9	21	SAT	ITS ACADEMIC	10	10			9	18	15	21	20	22	47	72	54	12	24	18	16	17	57	9	16	16	15	13	9
28	46	SUN	60 MINUTES	26	29	27		27	43	38	39	40	32	61	168	142	38	65	50	67	42	31	38	66	54	54	12
14	27	WCCC MON	TO TELL TRUTH	13	16	12	15	14	26	32	28	29	29	54	94	83	16	24	16	24	17	58	12	21	17	24	13	10
14	27	TUE	TO TELL TRUTH	16	12	13	15	14	27	30	28	29	26	52	95	85	11	26	20	30	22	52	8	16	14	19	13	9
14	27	WED	TO TELL TRUTH	14	18	16	14	14	27	28	27	27	27	51	93	75	11	22	13	21	22	51	8	14	14	18	18	14
13	26	THU	TO TELL TRUTH	14	15	14	14	14	26	27	26	27	27	49	87	70	11	18	13	19	15	57	13	15	15	18	18	17
13	26	FRI	TO TELL TRUTH	12	15	14	14	14	26	29	28	28	26	49	88	81	13	19	15	23	23	48	13	22	14	15	25	12
7	17	*AVS*	*TO TELL TRUTH*	*13*	*14*	*14*		*14*	*27*	*27*	*27*	*28*	*27*	*51*	*97*	*78*	*12*	*20*	*12*	*22*	*19*	*53*	*10*	*17*	*14*	*22*	*14*	*12*
7	17	SAT	EYEWIT NWS SAE	10	6			8	16	9	6	12	19	58	58	35	12	17	12	12	8	40	8	9	18	15	4	4
		NOR	*EYEWIT NWS SAE*	*6*	*6*	*13*		*8*	*16*	*18*	*11*	*12*	*15*	*47*	*57*	*37*	*11*	*17*	*12*	*13*	*6*	*47*	*6*	*8*		*22*	*4*	*4*
16	27	SUN	DREW&HARDY BOY	12	10	20		15	24	28	27	26	27	61	91	69	27	50	40	44	20	49	20	34	25	29	57	40

A total of 94,000 households viewed Monday's telecast of *To Tell the Truth.* These households may have been from outside the DMA as well as within it. In those households, 83,000 women 18+ were watching, 17,000 women 18-34, etc.

The station's share of audience in that time period during the previous measurement cycle was 32. An "x" in the column indicates that programming during the previous all-market measurement period was the same as in the current report.

		Program													
7.30P	14 25	WAAAMON HLLYWD SQUARES	13 13	14	16 14	25 20	23X 20	29 23	56 56	90					
	19 34	TUE MUPPETS	22 15	11	18 21	32 28	28 30	29 28	56 56	120					
	14 27	WED ALL-FAMILY M-F	14 11	14	14 13	23 21	20 18	23 20	54 54	82					
	11 20	THU HLLYWD SQUARES	11 10	11	10 11	20 16	16 22X	22 21	54 54	86					
	15 29	FRI NAME THAT TUNE	15 15	16	13 13	24 24	22 19	25 23	55 55	86					
	12 22	*AVG HLLYWD SQUARES*	*13 11*	*13*	*13 13*	*22 21*	*19 19*	*23 22*	*55 55*	*77*					
	16 29	W							56 56	104					
	14 25	MON							58 58	82					
	14 25	TUE							58 58	103					
	13 23	WED							54 54	88					
	12 19	THU							60 60	75					
	15 29	FRI							59 59	97					
	12 19								60 60	77					
	14 26								59 59	98					
	12 28								48 48	78					
	15 27	WCCC MON $128000 QUESTN	14 15	12	13 23	25 24X	24 24X	56 56	87						
	13 24	TUE ALL STAR-GOES	21 16	11	12 21	27 23	23 23	56 56	78						
	18 32	WED GONG SHW	21 16	10	16 30	33 33	30 20	54 54	100						
	13 35	THU FAMILY FEUD PM	18 18	16	13 26	26 18	21 17	54 54	104						
	12 25	FRI THATS HLLYWOOD	12 13	12	16 13	16 12	18 12	51 51	84						
	3 7	SAT KIRKLANDACO	3 3	3	3 5	5 9	9 9	48 48	18						
	3 7	*NOR KIRKLMNDACO*	*2 3*	*3*	*3 6*	*6 9*	*9 9*	*48 48*	*19*						
	<	SAT WLD-WRLD-ANMLS	3	5	5 10	10		48 48	29						
	9 16	WDDD MON BEWITCHED 1	6 8	5	7 12	13 8	8 13	56 56	44						
	7 12	TUE BEWITCHED 1	6 7	5	9 11	11 9	10 9	56 56	33						
	7 14	WED BEWITCHED 1	5 5	4	6 11	11 11	11 11	54 54	37						
	8 14	THU BEWITCHED 1	7 7	4	5 11	11 10	11 10	54 54	36						
	8 14	*FRI BEWITCHED 1*	*7 7*	*5*	*6 11*	*11 11*	*11 11*	*54 54*	*38*						
	6 13	SAT BEWITCHED 2	4 4	3	4 9	9 7	14 14	48 48	30						
8.00P	21 33	WAAA MON HOUSE-PRAIRIE	22 22	18	24 34	34 26	26 26	64 64	138						
	27 33	*NOR HOUSE-PRAIRIE*	*22 22*	*18*	*29 30*	*30 33*	*34 33*	*64 64*	*138*						
	<	TUE AMERICA-QUEEN		16	23	22	29	69 69	95						
	16 26	TUE MAN-ATLANTIS	11 23	8	14 22	22 21	24 20	63 63	90						

Fig. 8-9. Sample of information provided by the Nielsen station index (courtesy A. C. Nielsen Co.).

- limited financial resources associated with the economic depression of that time period,
- competition from commercial broadcasters for the available spectrum space,
- educational short-sightedness, and
- a lack of governmental commitment to the use of radio for education.

A point of view existed that the need for educational programming could be met by the use of air time on the commercial stations. During the development of the Communications Act of 1934, the need to reserve spectrum space for noncommercial use was discussed, but not enacted. It wasn't until 1938 that any such action was taken. In 1945, 20 FM channels were allocated to noncommercial use. Educational stations were authorized by the FCC in 1948 to operate on low power, as low as 10 watts. In the development of noncommercial FM radio during the late 1940s low-power stations focused on specific, localized needs and did not readily organize into a strong national force.

Television

Noncommercial television can trace its beginnings to the early 1930s when universities such as the State University of Iowa, Kansas State College, and Purdue University began experimenting with television. The growth of noncommercial television was slow. There was no great demand from the public at large for this system of broadcasting. As happened with noncommercial radio, one school of thought contended that the needs of education for television instruction could best be met by commercial stations who, after all, were already equipped with all necessary facilities and whose staff possessed production expertise. Due to the lack of their own stations, universities did, in fact, use commercial stations.

Public, governmental, and even educational apathy toward noncommercial broadcasting was illustrated in the thinking of the FCC when it revised its table of station allocations. The rapid growth of television forced the FCC to review and revise its plan for allocating spectrum space so that a comprehensive television service could be provided for the entire country. The commission decided in 1948 to stop issuing new license until its plan was developed.

The FCC deliberated the matter. Initial proposals made no provision for reserving channels for the exclusive use of educational television (as noncommercial television was then known). Of the seven FCC commissioners, one—Frieda Hennock—contended that educational channel should be established. Legal precedent for such an action had been established when the commission had reserved spectrum space for educational radio.

The battle for the rights of educational television was on. Through the efforts of educational leaders, committees and associations concerned with educational broadcasting, and Commissioner Hennock, the FCC revised its position. When the freeze was lifted in 1952, 242 television

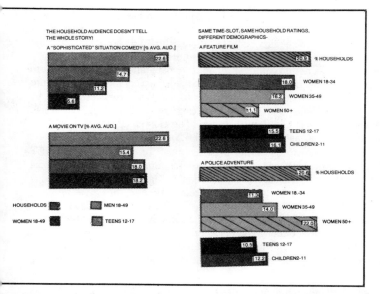

Fig. 8-10. Typical audience demographics for various program types aired during the same time period (courtesy A. C. Nielsen Co.).

channels were set aside for noncommercial television. (Today, there are 260 noncommercial television stations and approximately 1,000 noncommercial radio stations.) See Fig. 8-11.

This victory was just a beginning. The FCC's action was only a recognition of the need for educational stations. No provision was made for either the financial or administrative elements necessary to put the empty channels on the air. This meant that the activation of the newly assigned educational frequencies would be a local responsibility, a responsibility that had to be met without delay. Educators knew that failure to put the reserved channels into operation could result in the loss of the frequencies to commercial broadcasters. It was a matter of use it or lose it.

Had it not been for the Ford Foundation, the struggle could have been lost. This philanthropic organization poured hundreds of millions of dollars into the growth of educational television, providing the lifeblood of many stations. Eventually, stations received financial assistance from the federal government. The 1958 National Defense Education Act provided funds that stations could use for the production of in-school programming. Congress passed the Educational Facilities Act of 1962, granting money for both the activation of new stations and the improvement of existing facilities.

Most of the channels the FCC reserved for educational broadcasting were in the ultra-high frequency (UHF) spectrum. However, many home receivers could not even receive UHF signals. It wasn't until 1964 that manufacturers of television sets were required to produce sets that would receive both UHF and VHF signals.

Funding continued to be a concern during the 1950s and early 1960s. Noncommercial stations were forced to operate on tight resources. Program production quality reflected this austerity. Program distribution was a problem since stations were not interconnected into an electronic network. It was not possible for all educational stations in the country to transmit the same program at the same time from a common source to the entire country. (The first live, coast-to-coast network broadcast by noncommercial television wasn't braodcast until 1967). Programs were, instead, mailed from station to station. The closest approximation to an educational broadcasting network was the National Educational and Radio Center, later known as National Educational Television (NET). NET provided, via the mail, a much needed programming service to educational stations. Even this assistance was not enough. Many stations were "black" (not on the air) on weekends and during certain times of the day.

In 1964, the National Association of Educational Broadcasters spearheaded the drive that established a commission to study the needs, especially funding, of noncommercial television. The commission, known as the Carnegie Commission, released its report in January, 1967, citing the need for a well-financed interconnected educational television system.

The Public Broadcasting Act of 1967

The recommendations of the Carnegie Commission were given reality in the form of the Public Broadcasting Act of 1967. This landmark legislation created the Corporation for Public Broadcasting (CPB). In addition, the Educational Facilities Act was extended for three years. The Public Broadcasting Act was a commitment of the federal government to assume a responsibility for funding noncommercial broadcasting. It was the first step toward the establishment of long-range financing that would enable noncommercial broadcasting to plan its future. Nearly ten years later, the Public Broadcast Financing Act of 1975 provided authorization for 5-year financing. A ceiling of $88 million dollars was authorized for 1976 and was to increase to $160 million in 1980. The Public Telecommunications Financing Act of 1978 further authorized levels of $180 million in 1980, $200 million in 1981, and $220 million in 1983.

The Corporation for Public Broadcasting

The Corporation for Public Broadcasting was given the mandate to assist in the development of new stations, support programming at both local and national levels, obtain grants, and develop an interconnection system. In the early stages of development, the CPB was the administrative organization for the redistribution of funds provided to noncommercial broadcasting by Congress. It awarded grants to local stations and to independent production units. The funds provided by Congress are granted on a matching basis. Local stations must generate their own funds from donations, public and private sector contributions, and station memberships. For every two dollars raised by public broadcasting, Congress provides one federal dollar.

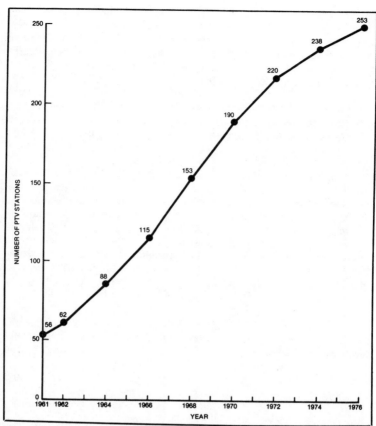

Fig. 8-11. Curve showing the increase in public television stations between 1961 and 1976 (courtesy Corporation for Public Television).

Although CPB funding is provided by the United States Congress and the corporation's 15-member board is appointed by the President (with the advice and consent of the senate), the CPB is not an agency of the government. An obvious danger in having a government funded and *controlled* broadcasting system is that the political party in power could attempt to use programming to further its ideology, to perpetuate its power, and to cover up its misdeeds. In addition to making the CPB independent of direct governmental control, board membership is designed to minimize the influence of a particular political party. (The United States has two dominant political parties.) No more than eight of the 15 CPB board members may belong to the same party.

Additional assurances of the CPB's political integrity and independence is rooted in certain restrictions placed by the Public Broadcasting Act. The CPB cannot own any stations. Nor can it produce its own programs (it has to be subcontracted). In fact, the corporation does not even schedule the programs that are shown on the public network. This function is the responsibility of the Public Broadcasting Service (PBS). The CPB created the Public Broadcasting Service to develop the network interconnection. PBS also does not own any stations or produce its own programming. Although established by the CPB, the PBS is controlled by the stations.

As an independent organization, PBS is free from direct political and governmental control. It has scheduled public affairs programs that were sometimes critical of the power structure. As could be expected, certain sectors of government took issue with such programs. Some observers believe that political intervention was attempted.

Conflict developed between the CPB and the PBS over the role each was to play in determining programming for the network. CPB took the position that since it provided funding and since it created PS, it should determine programming. PBS should limit itself to the mechanics of network operation. Of course, PBS disagreed, arguing that CPB should not be a programmer, but an advocate of public broadcasting, a solicitor of funds, and a protective barrier between public broadcasting and political pressure. CPB threatened to exercise programming control by ceasing to fund those programs to which it objected. Eventually, a compromise was made and a partnership of sorts was created in terms of programming.

As a result of the conflict, PBS revised its structure so that more control and financial support was given to the member stations. Programming decisions for the public network would center around the needs of the local stations. The Station Program Cooperative (SPC) was established as a means of acquiring programs. The SPC analyses the needs and interests of both the stations and the public. Program proposals are then solicited from stations. The stations review the proposals and indicate the ones they are willing to "purchase." In this instance, "purchase" means that a station agrees to pay a share of the production costs.

PBS programs are distributed to stations via a satellite interconnection system (Figs. 8-12 and 8-13).

National Public Radio

In 1971, the CPB created National Public Radio as a network for noncommercial radio. During the early years of economic struggle, noncommercial radio was not electronically interconnected. National programming was syndicated by means of an audio tape mail network. Radio development was hindered when noncommercial television caught the public's fancy and, as a result, noncommercial radio continued to struggle while television progressed.

The establishment of NPR did not mean that every noncommercial radio station was automatically a member of this new organization. NPR established certain requirements (staff, hours of operation, and budget) that had to be met in order to qualify for membership. The intent was to build and support a system of public radio stations that would develop into a strong national service. NPR wanted to avoid perpetuating the fragmented, weak nature of noncommercial radio as personified by the low-power stations. Only about 18% of noncommercial radio stations meet the qualifications. Low-power stations are often an electronic "junkbox" operated essentially as a college club. And, universities with academic studies in broadcasting use these stations as a training facility.

A strong nationat impact was necessary for public radio to ever fully develop and to realize its potential. NPR has yet to make its desired impact. According to a 1978 study, only 28% of those surveyed had ever heard of NPR. As low as this may seem, it's much better than the 11% recognition factor indicated in a 1976 survey. Whereas PBS does not produce its own programs, granting funds for this purpose to stations, NPR not only provides station grants, but it also produces programs.

The network capability of public broadcasting has made incredible advances within the past few years. From a tape-by-mail network, public broadcasting is now interconnected by satellite and has technologically surpassed commercial radio. The programming potential offered by satellite has had an impact on the organizational structure of PBS. With three channels of programming being available, PBS can now offer three different program feeds, including a prime-time service, an instructional service, and a channel devoted to regional and special interest programming. In effect, PBS will be providing three different, independent networks. However, all three services have been placed under the central authority of a programming president. As a part of its reorganization, PBS divided its functions into more discrete areas. The Center for Public Television was created to function as a trade association.

Fig. 8-12. The above drawing illustrates how programs are distributed over public television's new satellite interconnection system. Programs are transmitted to the WESTAR communications satellite from PBS's main origination points around the country. The satellite amplifies and re-transmits the programs for reception at ground terminals serving each public television station licensee (courtesy Public Broadcasting Service).

Noncommercial Broadcasting as Public Broadcasting

The emergence of public broadcasting from what was educational broadcasting has involved a process of redefinition of purpose and of programming philosophy. The term *educational* was found to be too restrictive. Although most stations, especially those operated by schools and universities, do provide programs intended to meet educational and instructional needs, programming is also devoted to the fine arts and to public affairs.

Noncommercial as a descriptor is too vague, saying little more than without commercials. The term *public* is subject to different interpretations. Some stations are public because they are financed by public tax dollars, such as those stations operated by local school boards and those licensed to a state commission or authority.

Another type of public station is the *community* station. These stations are licensed to many nonprofit corporations having a board of directors representing various special interest groups (labor, business, education, fine arts, etc.) in the community. Individuals within the community may buy a station membership by paying a yearly contribution of approximately $15. Membership in the community station usually includes the right to vote for the governing board. In a sense, the public has a voice in deciding station governance.

Public broadcasting, according to an often used phrase in the Carnegie Commission Report, should be the "bedrock of localism." For some, this phrase implies that the programming of public broadcasting shoud center around local, public needs and wants. Through community-oriented programming, public broadcasting can keep citizens informed of the actions of local government, and important issues affecting a community can be bought into focus. An extension of this thought implies that public stations should serve as the voice and eyes of the community.

NEW TECHNOLOGY

The limited spectrum space has made the business of broadcasting something of a government-approved monopoly. Technical limitations have, in the past, bound audiences to stations. Network-affiliated stations, in turn, have been tied to their respective networks by an umbilical cord consisting of land lines and microwave links. The use of satellites for program distribution has dramatically changed these old relationships. The implications are exciting for the public. However, commercial broadcasters may have feelings of uncertainty.

It was announced in 1979 that certain syndicated programs would be distributed to commercial stations by means of satellite transmission. Reportedly, this method would be more efficient than the old method of mailing the programs to the stations and then having the stations return them by the same method. Each participating station would have its own receiving antenna or "dish." The Southern Baptist Radio and Television Commission has found satellite transmission a most economical means of

Fig. 8-13. The PBS main origination terminal, near the Washington suburb of Springfield, Virginia, is the hub of public television's new satellite system. From here, up to four programs are simultaneously transmitted to Western Union's WESTAR I domestic communications satellite, orbiting 22,300 miles above earth. WESTAR I then re-transmits the programs to public television stations in the continental U.S., Alaska, Hawaii, Puerto Rico and the Virgin Islands (courtesy Public Broadcasting Service).

delivering public-affairs programming. Traditional methods cost the commission $126,000 a year to service 800 cable systems. By using the satellite, the cost is just $25,000.

Operationally, the use of satellites for signal transmission presents no significant problems for the local station. The signal is received by the "dish" and routed to the station for redistribution. A station would have the option of either instantly retransmitting the incoming signal or of videotape recording it for delayed playback.

In effect, a mininetwork is being created between the program originator and the participating stations. Producers and creators who are unable to get their programs placed on the broadcast schedule of either ABC, CBS, or NBC will now have an opportunity to go directly to stations via satellite. It may be a matter of time until network prime-time programming is eroded by satellite based mininetworks.

CABLE TELEVISION

Cable television systems, by using satellite transmissions, have already emerged into a television "network." Known originally as community antenna television (CATV), cable television began in areas that were unable to receive a television signal of good strength and quality. Cabled communities are usually either some distance from a broadcast station or are surrounded by hills or mountains that block reception. Tall buildings in large cities can also cause poor reception.

To overcome this poor reception problem, cable operators place a very tall tower at the top of a site with a high geographic elevation, usually a hill. Antennas for receiving signals from several stations are attached to the tower. Incoming signals are amplified, processed, and redistributed by coaxial cable to participating homes (Fig. 8-14). Homes using the cable service are charged a monthly fee of approximately $8.

Even some homes capable of receiving good signals with their home antennas subscribe to cable television. The essential characteristic, making cable attractive even when not needed for good signal reception is the multitude of channels offered on a cable system. For example, homes in a medium-sized market receive signals from four stations (the three network affiliates and the public station). In contrast, a basic cable system can provide a minimum of 12 channels. Technical developments can make possible 24 or, theoretically, hundreds of different channels.

Restrictions

Cable operators are required by FCC regulations to carry the three local network affiliates and the public station. Additional stations can be imported from other communities and placed on the system. However, FCC regulations contain some restrictions on signal importation (See Table 8-1) designed to protect the local stations from the competition provided by the out-of-town stations. The number of stations that can be imported depends on the size of the market in which the cable company operates.

Cable operators are also required to provide *nonduplication* protection of the local station's network programming. If a local station carries a network program, that same program must be deleted (blacked out) on any imported stations. Regulations have also been developed in regard to the exclusive rights of a local station to the syndicated programs it purchases.

Program Origination

Since the number of distant stations a cable system can import is restricted, cable operators have sought other programming for their many

170

channels. Most systems dedicate one channel for the display of time and weather data. Another channel may provide a news service. At a more elaborate level, a cable system may have a studio and produce its own programs.

Cable productions are often similar to programs produced at broadcast stations, e.g., talk-interviews, news, and sports events. The technical standards required by the FCC are not nearly as strict as those required for broadcast stations. Therefore, cable programs can be produced with low-cost, light-weight, industrial quality equipment.

The combination of abundant channel capacity and less stringent equipment standards led early advocates of cablecasting to be overly optimistic and to make excessive claims for the potential of locally-produced cable programming. It seemed so logical that cable, with its many channels, could become the "eyes and ears" of the public, a role that was once expected of public broadcasting. Channels could be dedicated to the special needs of government, public expression and education.

Many schools and colleges had (and still have) closed-circuit equipment systems that were compatible with equipment used by cable television. Educational programming would, so it appeared, by only a matter of either letting an institution serve as the origination point for one of the cable channels or of providing prerecorded programs to be played from the cable system's own signal origination center (known as the head end). Not only could instruction be provided for in-school use, but the reach of education could be extended. Cable television was seen as a vital

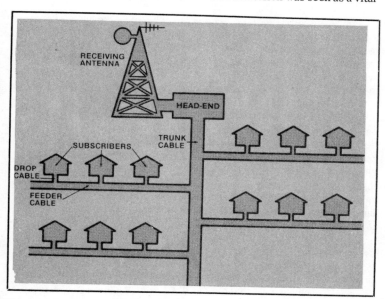

Fig. 8-14. Block diagram of a basic cable TV system (courtesy National Cable Television Assoc.)

element in the development of nontraditional learning, of creating "schools-without-walls," of making learning a life-long process.

Cable channels devoted to government access were seen as a means of keeping the public better informed. Meetings of local government could be telecast and the debates and discussions on critical issues would be brought under closer public scrunity. Public-access channels would provide individuals with an opportunity for creative expression and for social-political commentary. The great potential of cable to provide programming to meet the needs of specific audiences was conceptualized as *narrowcasting*.

For a variety of reasons, cable requirements for local production and for public access were modified or eliminated. Educators found that the equipment, staff, and funding required to provide a regular schedule of programming to be a strain on already limited financial resources. Cable is regulated at the federal, state, and local level. FCC regulations generally regulate to technical standards, cablecasting services, signal carriage, and distribution of regulatory authority. State and local government may regulate in those areas not expressly under the authority of the federal government. Local government usually requires that a cable company secure a franchise before it will be permitted to operate in a community. The franchise agreement will establish the rate that the operator can charge a subscriber. The cable operator may be charged a fee, usually a percentage of gross annual revenue, for the privilege of being granted the franchise. Public services may be specified in the agreement, such as free services to schools or public access.

In recent years, *local origination* has taken on a meaning other than local production. Many systems now feature programs produced by or received from outside sources. Institutions seeking broader public relations impact provide cable operators with free programs. Cable systems also buy programs *packages* (primarily syndicated programs, movies, and sports events) that are designed to attract a mass audience.

As a business operated to make a profit, cable systems must decide how the costs of program rental are to be absorbed or recovered. On the one hand, program origination can be considered as an investment designed to attract new subscribers to the system, especially in areas with acceptable over-the-air reception. An adequate base of subscribers is essential if a new system is to be economically feasible, since the line installation charges are enormous. The average cost per mile of laying cable is $6,000 in rural areas and $10,000 in cities. If the cable has to be laid underground in a city, the average cost can go as high as $100,000 per mile.

Rented programs can also be a source of additional income. Some systems sell advertising time. Perhaps the most significant development is pay cable, a procedure whereby the customer is charged a fee for the programs he watches. A common technique is to transmit pay programs on a special frequency that cannot be received without using a special piece of equipment that is attached to the viewer's television set.

Pay cable is an optional service that costs subscribers nearly $8 a month. This is an addition to the basic service charge. The monthly charge

Table 8-1. Cable TV Signal Importation Restrictions.

Market Size	Number of Independent Stations That Can Be Imported	Bonus Stations That Can Be Imported
Top 50	3	2
Second 50	2	2
Below Top 100	1	none
Source: FCC		

does provide the subscriber with comprehensive access to all pay cable programs transmitted in a given month. Some systems, however, charge on a per program basis. The popularity of pay cable has grown to the point that more than 900 cable systems are providing pay cable services to nearly 3.3 million subscribers.

The development of pay cable has been significantly advanced by the use of satellites for program distribution. By early 1978, the FCC had received 1525 applications for earth-receive stations. Nearly 1350 of these were from cable companies. Program suppliers such as Home Box Office, Showtime, and others can now transmit their programs by satellite to affiliated cable systems throughout the entire country. The interconnection of cable systems by satellite has, in effect, created another television network.

Why has pay cable become so popular? Why is the American public willing to pay for television programs when free programs are being provided by broadcast stations? A part of the appeal is the lack of commercial interruptions. The attitude that commercials are a fair exchange for free programs is apparently changing for some Americans. Another major attraction of pay cable is the opportunity to view fairly recent movies that have not gone through the editing and censorship (usually of profanity and nudity) common to commercial television. Movies, in fact, constitute over half of the programming provided by two of the major suppliers of pay cable programming. Cable systems can be more daring in the content they present, since they do not need a license from the FCC to operate.

From the broadcaster's point of view, cable television is an economic threat. The importation of stations from other markets provides additional competition for the attention of the local audience. A loss of viewers to an imported station could affect the value of a station's commercial time. Cable systems counter with the argument that a station's coverage is increased whenever its signal is redistributed into new areas by cable. Broadcasters have lately been concerned about a particular type of imported station, the "superstation." Whereas most imported stations are from the same geographic region, superstations are transmitted by satellite to cable systems throughout the entire country (Fig. 8-15). One such station, Atlanta's WTCG, reaches nearly 3.3 million homes.

Broadcasters are also concerned about the impact of cable-originated programming on audience fragmentation. Will viewers turn from broadcast stations in preference for cable programs? For the moment, the issue is not yet critical. Cable television, with 4100 systems in 9000 communities, has only about 14.5 million subscribers, or 20% of the total television households in America. (There were only 70 communities with cable in 1950). Total cable revenue is over $1 billion a year.

Cable's economic strength will grow as cable penetration increases. Within the next five years, cable should achieve 30% penetration level. In numbers, this means over seven million new subscribers. As the financial resources of cable increase, broadcasters will also be faced with competition for the rights to quality programs and events. If cable is able to compete on equal terms with broadcasters for program rights, either audience fragmentation will increase or the law of supply and demand will increase the cost of programming.

Future developments in broadcasting and in cablecasting will depend on several factors. Capital for cable penetration into urban areas may become available as public demand for pay cable services grows. Government regulation will play a critical role. Will the FCC rules favor and protect broadcast stations? Or, will broadcasting be at the mercy of the marketplace? The conflict between cable and broadcasting may be an insignificant point if satellite transmission of programs directly into the home ever becomes a practical reality.

The distribution of programs to the home by satellite has serious implications for stations and networks. If the majority of a local station's programming comes from outside sources (networks and syndicators) and if satellites can provide that some programming—and more—through cable systems or a direct feed, why should the local station continue to exist? By eliminating the middle man (the local station) could not programs be supplied more economically and efficiently?

There is no clear and easy answer to these questions at the moment. Certainly, the local station manager would answer "No!" arguing that the local station serves several vital roles. Through public affairs and news programs, the local station provides an essential service to the public interest. Without these programs, the public would be less informed of critical issues in their community. An important forum for the expression of opinion would be lost. The local station also serves as a buffer or contact between the public and the program suppliers. Should a program violate community standards, citizens can express their displeasure to the station, pressuring it to either not broadcast the program series or to, in turn, apply pressure to the network.

HOME VIDEO

While broadcasters are keeping their corporate eyes on the satellites flying over the homes of their audience, another development is occuring within the homes themselves. American consumers have been attracted to the home videocassette recorder (VCR) and, more recently, to the video disc.

Videocassette Recorders

Videocassette recorders are promoted by their manufacturers as a *time-shift* machine to record programs off the air for viewing at another time. Typically, VCRs are used when factors such as sleep, work, and interest in another program being presented at the same time causes a viewing conflict. The VCR has an internal tuner for channel selection and a timer (optional) that automatically starts and stops the recording process. In addition, a portable camera can be purchased for use with the recorder. (See Fig. 8-16.)

Video Discs

Video discs, on the other hand, do not have a record function and can only play back prerecorded materials. Although the lack of recording capability may appear to be a marketing disadvantage, the video disc is competitive in price. A VCR costs approximately $1,000 and its programs are in the $40 to $80 range. Costs for video discs, both players and programs are about half as much.

Home video recording of broadcast programs presents a few problems for the broadcast industry. Audience measurement is more

Fig. 8-15. The RCA Sat-Com satellite has been used extensively for cable programming (courtesy RCA Americom).

complicated. By using the recorder's pause or fast-forward controls, commercials can be avoided. Penetration of home video is expected to reach at least 10% during the 1980s. Even then, any significant effect would require that extensive off-air recording take place.

Prerecorded Program Sources

The greatest significance of home video will most likely come from the use of prerecorded materials. Past development of home video has been caught in a circular trap. Programs weren't available in large numbers because machines weren't in widespread use. Machines weren't purchased because programming was limited. Now, programs and their suppliers are established.

The consumer can select from a wide range of programs, including educational, sports, hobbies, movies, old television shows, "X-rated, sex-oriented" features and music. In fact, music programming could become a big selling category. Inexpensive video records featuring top performers in concert could be as popular with American teenagers as phonograph records are today. Program-of-the-month clubs are another possibility. And, friends could always exchange programs and thereby reduce costs. Program exchange makes for an interesting mathematical analysis. If 20 friends each bought one program at $20, an exchange would reduce the per program cost to $1 per program!

The fact that consumers can now buy exactly the types of program needed could impact both broadcasting and cable television, possibly fragmenting the audience. Home video may even emerge as a competitor for programming. If program suppliers can deal directly with the home consumer at a competitive price, the attractiveness of pay cable may diminish. Even public broadcasting could be affected.

Those seeking cultural programming may decide to purchase tapes or discs featuring a wide range of the fine arts. If education is a goal, why not purchase one's own instructional programs? The learner could progress at his own rate and the programs could be reviewed when difficult material is hard to understand. Rather than being perceived as a threat by public television, home video may be understood, within the public telecommunications center concept, as only one of several alternate delivery systems.

Video recorders and video discs are connected to the home television set for program viewing, making the home a closed video system. In a sense, the American public has achieved a certain independence from the broadcaster. This is just one of several challenges broadcasting must face in the future.

THE FUTURE: REGULATION VS DEREGULATION

While this chapter was being written, the United States Congress was considering proposals that would rewrite the Communications Act of 1934. The legislation, if passed, could greatly deregulate broadcasting. Requirements such as the fairness doctrine, the equal-time provision, and the community ascertainment survey would be deleted. The 3-year

Fig. 8-16. Home video recording equipment (courtesy Sony Corporation of America).

licensing period would be eliminated. Radio would be licensed for an unlimited time period, as would television, after providing 10 years of service. The public interest standard would give way to the forces of the market place. Only if these market forces failed would the government intervene.

Revision of the Communications Act is seen as a necessity partly because of the progress made by cable and satellite technology. This new competition may have a significant effect on broadcasting's economic base and revolutionize the industry. At the very least, the monopoly that broadcasters once had on program distribution is over.

As can be expected, the proposed deregulation has been a source of debate and controversy. Those favoring deregulation assert that market place competition will indeed make broadcasting, as a matter of good business practices, responsive to the public interest. Failure to serve the community would result in audience turning to those stations who did. In a sense, the public interest would be regulated by the public itself as it selects or rejects a station. Repeal of the fairness doctrine and the equal-time provision would grant stations increased flexibility to pursue public affairs programs without concern for governmental criticism.

Opponents of deregulation argue that the need to maintain open and fair communications is too important to trust to voluntary self-regulation, that deregulation will further increase the control and power of the existing communications monopoly (nearly a third of cable television is owned by corporations having an interest in broadcasting), and that the concept of a competitive broadcasting market place is not valid.

177

Passage of the rewrite legislation will bring many uncertainties with it. Most importantly, can broadcasters be trusted to serve the public interest? Will the voice of the people be strong enough to truly effect change in the programming process? If not, what means will be available for a citizen to challenge an abuse of power?

Ironically enough, each side of the deregulation issue uses the First Amendment's right of freedom of speech and expression as a justification for its point of view. It may be a question of who's First Amendment rights are greater: the public's right for an open exchange of ideas, as mandated by regulation, or that of broadcasting to have the same right of expression as that granted to other media.

In the final analysis, the issue of broadcast regulation, of deciding who shall determine and protect the public interest, is a matter of the prevailing philosophy of governance held by a society or culture. An individual point of view is shaped by one's own politics, and conscience. To deregulate broadcasting will be an act of trust and faith: trust in the goodwill of the broadcast industry and faith in the wisdom of the people and the safeguards inherent in a democratic society.

Chapter 9
Broadcasting In Brazil

Esmeralda Eudoxia Goncalves Teixeira
Departmento Nacional De Telecommunicacoes

A clear understanding of the rapid development of telecommunications in Brazil demands a brief presentation of some facts regarding the nation's history and the difficulties which had to be overcome on the path to our current era. The country's territory is of continental dimensions, measuring 8,500,000 square kilometers, divided into 23 states, four territories, and the federal district, the area in which the capital of the republic is located.

NATIONAL HISTORICAL BACKGROUND

During a period of more than 300 years after discovery, 1500 to 1808, Brazil, which was nothing more than a Portuguese colony in the Americas, saw its primary riches, such as gold, diamonds, precious stones and *pau-brasil* (a valuable type of wood), gradually devastated. In return, the new land received nothing, not even the minimum conditions demanded by progress.

Many were the invaders which attempted to make their way ashore. Interested in forming colonies here were several French invasions, followed by the English and the Dutch; but, all were repelled by the coastal defenders. Only the Portuguese remained.

The general governors, who were nominated for Province of Brazil by the King of Portugal, sponsored expeditions into the interior in search of natural riches. These groups, called *Entradas*, and the *Bandeirantes*, whose expeditions had the same objective and were known as *Bandeiras*, together with the Jesuits who came here to covert the indians, were those principally responsible for the expansion of the country's boundries which, according to the terms of the Treaty of Tordesillas, were to take in less than half of the present area.

Besides the Bandeirantes, there were few who had the courage to make their way into the interior. The first population centers were scattered along the coast and were called *Eeitorias*. Later, the hereditary captaincies were created, and these, in turn, gave rise to the present states and cities.

Since it was in the interest of the Portuguese to maintain Brazil as a permanent colony, little was permitted in the way of social, commercial or industrial development.

Few Brazilians could read or write, and measures were taken by the Crown to restrict the access which the literate population had to books. A number of individuals, in the hope of creating rudimentary system of communications, tried to open small printing shops, but these efforts were soon frustrated.

The people were kept in a marginal position as regards the development of contemporary currents of thought, for, if reading was improbable, the printing of pamphlets or similar types of literature was impossible.

Even in the Court, the press was subject to repressions and censorship. However, a highly important historical event brought a glimmer of hope to those who desired progress. Since King Dom Joao VI refused to take part in the blockade aimed at depriving England of its trade with the continent, the Emperor of France, Napoleon Bonaparte, fulfilled his threat of invading Portugal, forcing the Court to take up residence in Brazil.

Here in Brazil, the King had to take measures to provide the members of the Court with improved living conditions, while making it possible for the government to operate in a reasonable fashion. The arrival of the royal family in Rio de Janeiro led to the first steps in the direction of progress: the ports were opened to friendly nations; Brazil was raised to the rank of a Kingdom United to Portugal and Algarves; the Bank of Brazil was founded and the Royal Press, subject to censorship and royal prerogatives, was permitted to function.

Later, a number of printing offices were allowed to operate in Bahia and Pernambuco. Although they soon began printing newspapers and pamphlets, they were soon closed down by the censors, who tended to be quite severe as regards the printed news. During the same period, the Royal Post was founded and consisted of riders who travelled from one province to another carrying orders and messages.

Rising political tumult in Portugal made it necessary for the King to return and, on April 24, 1821, Dom Joao VI left for Lisbon, leaving behind his son, Prince Dom Pedro, as regent of Brazil. The young prince was later to liberate Brazil from the yoke of Portugal and assume the office of the emperor.

EARLY TELECOMMUNICATIONS

During the reign of Emperor Dom Pedro I, little took place in the way of economic and social progress, for it was a period of roganization and the creation of the essential structures. However, when Dom Pedro I passed authority on to his own son Dom Pedro II, he left Brazil in a state of preparation for the challenges of the future, ready to assume its position in the community of nations.

During the second kingdom, many improvements were introduced to Brazil, including a number directly related to the area of telecommunications. The first electric telegraph line was inaugurated on May 11, 1852, while the telegraph center was placed under the responsibility of the ministry of justice. The necessity of organizing the telegraph service led to the elaboration of the first Brazilian Telecommunications Code. The original lines were installed for the sole purpose of serving the needs of the government and it was only in 1858 that public utilization was permitted.

On the initiative of the Baron of Maua, Brazil was connected to Europe by means of an underwater cable in 1872. However, due to the high operation cost, Brazilian firms were unable to care for the functioning of the international lines. The installation of the system demanded foreign capital and, due to this, the Western and Brazilian Telegraph Company began to operate in Brazil, providing capital support to the installation of the domestic and international telegraph system. Thus, during the Imperial Age, Brazil sought to make up for the time lost during the colonial period.

Brazil was one of the first nations in the world to make use of the telephone. It happens that Dom Pedro II was in the United States when that country was commemorating the Centenary of Independence. On that occasion, he visited the Centenary Exposition, held in Philadelphia in May and June of 1876, and had the opportunity of getting a first-hand look at an astounding new discovery called the telephone.

On returning to Rio de Janeiro, he ordered the Western and Brazilian Telegraph Company to install telephone lines in the palace located on the "Quinta da Boa Vista." The inauguration of the lines took place one year later, in 1877.

In the following years, a number of foreign companies received permission to install telephone lines joining different parts of the city of Rio de Janeiro to Niteroi. In terms of communications, the Imperial period came to a close with the telegraph, telephone and underwater cables in full operation.

With the implantation of the incredible inventions of Morse, Marconi, and Gutenberg, the immense distance between the different points of the vast Brazilian territory began to diminish gradually. During the same period, Hertz discovered the principle of electromagnetic wave propagation and made it possible for Brazil to enter into contact with the rest of the world.

THE BEGINNING OF BRAZILIAN RADIO BROADCASTING

In 1922, the people of Rio wildly celebrated that city's first long-distance sound transmission. The event took place during the first centenary of Independance, and consisted of a speech by the president of the Republic. The presidential speech reached the ears of the population in localities far from Rio, and was also heard over speakers in a great number of small interior towns. This was made possible as a consequence of the installation of a small transmitting station upon the heights of the *Corcovado* Mountains, which was the responsibility of the Westinghouse Electric International Co., and the Brazilian Telephone Company.

One year later, the Western Electric Company brought two 500-watt transmitters from the United States to be used in the telegraph service. One of these was made available to Brazilian amateur radio operators who, with the encouragement and leadership of Fransisco Roquete Pinto and Henrique Morize, opened Brazil's first radio station, the Radio Sociedade do Rio de Janeiro, on April 20, 1923. It was not long, however, before the second station opened in the city of Recife, capital of the State of Pernambuco. There, the Radio Clube de Pernambuco was inaugurated on October 17 of the same year.

After the inauguration of these two stations, others began to appear almost immediately in the large population centers along the Brazilian coast.

BRAZILIAN BROADCASTING LAWS

The first republican constitution modified the legislation governing the existent telecommunications services by granting a certain degree of autonomy to the states. However, just one year later, the federal government resolved to concentrate this authority in its own hands once again. In 1917, Executive Act No. 3,296, which dealt with the radiotelegraph and radio telephone services within the area taken in by the nation's boundaried and territorial waters, determined that authority over those services would be the exclusive responsibility of the federal government.

Due to the high costs involved, the operation of the telegraph services was delegated to foreign firms. Later, the licenses held by these companies were renewed only with regard to the telephone services, while the telegraph became the sole responsibility of the government, which operated it through the *reparticao Geral dos Telegrafos*.

In the case of radio stations, which did not demand such high investments, after the inauguration of the first two stations in 1923, enthusiasm for this new means of communications spread rapidly among the population. Since little in the way of capital was needed, societies were quickly formed and radio stations began to appear in all parts of the country. In 1924, various stations were inaugurated in the north, northeast, and the south and west of the nation.

Since a great number of stations opened in the large Brazilian cities, the government was forced to enact legislation governing this type of undertaking. Thus, Decree No. 21,111, issued on March 1, 1932, and known as the *Regulation for the Execution of Radio-communication Services*, brought together all the pertinent legislation in existence up to that time.

This decree, which contained 109 articles, was the last word in the field for a period of 30 years. However, during this time, the law did not go unchanged. A number of executive acts and regulations were introduced into the articles, a fact which eventually led to mutilation of this legislation.

Decree No. 21,111 defined the different types of radio-communication services, determined who was to grant licenses and operate in each specific field, set down the process to be followed in the granting of licenses, the suspension of the same and the penalties to be imposed in the cases of communication crimes, and created the Radio Technical Commission, which was granted administrative responsibility over the sector. It also determined questions regarding frequencies and determined that radio and television licenses would be granted for periods of 10 and 15 years and could be renewed for equal periods of time by the mutual agreement of the federal government and the operating entity.

With the passage of this decree, there began a process which witnessed an almost unending series of other laws and regulations. For the most part, these were contradictory in nature and, almost always, led to serious problems with regard to their application.

Since 1930, Brazil had lived under a dictatorial regime, which sought to limit and control both freedom of thought and the utilization of the communications media, particularly the written and spoken press. Thus, parallel to Decree 21,111, another law, called the press law, governed the freedom of thought. Obviously, both of these laws were in urgent need of reformulation.

The Present Telecommunications Code And Its Background

Between the publication of Decree 21,111 and its many alterations and the present legislation governing telecommunications, those operating in the sector waged a long and memorable, and sometimes violent, battle to unify the laws published since 1932, while adapting their tests to the needs of a changing world.

As a consequence of the first studies carried out for the purpose of setting up a single and unified body of legislation, a commission was created to deal with the subject in 1940. However, this effort was destined to failure, as was a similar movement which got underway in 1944. Due to the government policy in force (dictatorial regime of Getulio Vargas), censorship was severely imposed on both newspapers and radio. In other words, just about everything and everyone came under government control.

Obviously, those who labored in the sector were greatly dissatisfied. As a consequence, between April 8 and 13, 1946, individuals who worked in radio broadcasting from all parts of the country came together at the first Brazilian Radio Broadcasting Congress in Rio de Janeiro, for the purpose of participating in a decisive discussion of the problems suffered by the group. On that occasion and with the participation of representatives from the publish and private sectors, they drew up a proposal for the new Brazilian Telecommunications Code.

After discussing and voting upon the proposal, it was presented to the president of the Republic. Being the eighth proposal of its kind to be drawn since 1940, it came to the same end as the others—simply forgotten, without comment or analysis. Years of meetings and confrontations in the national congress were to no avail as the government refused to act upon the subject. Between 1947 and 1957, other proposals were presented.

It was only 22 years and 18 proposals later, after constant labor and discouragement, that the goal was finally reached. Finally, the national congress voted upon and passed the Brazilian Telecommunications Code. The bill was sent to the president who, to the surprise and indignation of all concerned, refused to sign it, vetoing 52 of the 129 articles.

The decision was received with astonishment, for vital articles of the law had been eliminated. Those who worked in the sector became suddenly aware of the necessity of joining forces, while mobilizing the opinion of the public and the members of the national congress, for only in this way would it be possible to override the presidential veto.

Within a few days, 172 representatives of radio and television stations arrived in Brasilia to meet with congressmen, clarify their doubts and, above all, demonstrate to them the immense promise of the new code in terms of the broadcasting sector. There could be no vetoes, no modifications or alterations which could mutilate 20 years of ardous labor.

Finally, the day arrived and the legal test, together with the vetoed articles, was placed before the national congress. The decision was awaited with great anxiety and, in the end, a unanimous decision was taken. Not one veto was upheld.

The new law went into effect in the precise form in which it had been presented. The results of this battle, the meeting of radio and television station owners during the congressional debates, led to the conclusion that what the union achieved should be made to last permanently. It was thus that the Brazilian Radio and Television Association (ABERT) was founded. Today, the Association has approximately 750 members.

The Brazilian Telecommunications Code - Law No. 4,117/62

Article 1 of this law states that telecommunication services within the country's borders, on its territorial waters and in its air space, as well as in those places to which international agreements and principles have granted extra-territorial standing, the precepts and regulations contained therein shall be obeyed. The law defines telecommunications services one by one, while it confirms that the federal government, according to constitutional principle, shall have the private right of directly operating these services, while supervising the services it permits others to provide.

Despite the fact that the operation of these services is the exclusive responsibility of the federal government, the federal constitution permits mixed operations, public and private sectors. As in many other sectors, it is the belief of the federal government that the presence of the private enterprise in this sector is not only beneficial but even decisive to the availability of higher quality services. With this conviction in mind, the government saw fit to place practically the entire system in the hands of the private sector and, with this decision, it has achieved excellent results.

In the former law, the penalties to be applied in cases of abuse, crimes or offenses were particularly severe. Together with a number of other articles, those provisions dealing with these penalties and their applications were revoked or rewritten by Executive Act No. 236, published on February 28, 1967. In the new law, a system of determining telecommunications rates was also defined.

On October 31, 1963, Decree No. 52,796 was signed into law approving, defining and classifying the Regulations of Radio Broadcasting Serivces. This decree became the handbook of those involved in the sector, for it includes just about every possible clarification of the subject along with such minimal details as models of petition forms.

In the early days of February 1967, the federal government issued Executive Act No. 200, creating the Ministry of Communications while carrying out a total administrative reform. The new ministry had the responsibility of proposing, supervising and carrying out the telecommunications policy of the new regime, which took power in 1964.

On February 28 of that same year, the already mentioned Executive Act No. 236 was issued for the purpose of modifying and adding a number of probisions of law No. 4, 117/62, so as to make it more compatible with the policies of the day.

The Brazilian Telecommunications Code, clarified by Decree No. 52,026 on May 20, 1963, reformulated the structure of CONTEL, creating the National Telecommunications Department as the executive secretariat organ, as well as the presidency of the council and the plenary organ (the deliberative branch). Today, DENTEL is the executive organ of the Ministry of Communications and the possessor of highly important attributions.

The validity of licenses remained the same as previously, 10 years for radio and 15 years for television, both of which are renewable for equal periods.

At that time, CONTEL, which is responsible for analyzing the juridical and technical situation of those stations seeking license renewal, was not sufficiently equipped to carry out its task. For this reason, law No. 4,117/62 permitted radio stations to continue their services until 1972, while television stations were allowed a period extending until 1977.

If one were to analyze the principal articles of the present code, he would find a profile of the broadcasting model as defined and adopted in Brazil. The text itself determines that the federal government may carry out broadcasting services "directly," while the government may give its permission to the following entities to carry out these services "indirectly":

- The States, Territories and municipalities;
- Brazilian universities;
- Foundations constituted in Brazil;
- Brazilian societies constituted by nominal stocks or quotas, with the condition that they be subscribed to by Brazilians.

Neither foreigners nor legal entities, with the exception of political parties, may be members of or participate in societies which carry out broadcasting services, nor may they exercise any type of direct or indirect control over them.

This prohibition does not include naturalized Brazilians, who may be members of these societies. However, the federal constitution expressly determines that the direction of these companies must be in the hands of native Brazilians.

Radio and television are particularly the only media which reach the entire Brazilian population, including all ages and social classes, having direct influence on the formation and behavior of the people.

For this reason, the government took care to avoid the concentration of the mass media in the hands of a restricted number of groups with specific commercial or ideological interests. In other words, the formation of a communications monopoly in Brazil was prohibited. Thus, each entity may be licensed to carry out broadcasting operations within the country, according to the following limits:

Radio Stations

Medium wave
from 250 to 500 watts,	4
from 1 to 10 kW,	3
Above 10 kW,	2

Frequency modulation,	6
Tropical waves,	3
Short wave,	2

Television Stations

10 in the entire country with a maximum of five in VHF and two per state

As a way of avoiding the formation of monopolies, the law determines that entities may not be granted licenses if individual holders of stocks or quotas in these companies are also stockholders in other broadcasting companies, beyond the limits presented above. These limits are applicable to legal entities, individuals, stockholders and directors.

According to law, educational services may be provided by television. However, by analogy, the Ministry of Communications has also extended this right to radio stations.

There is no doubt that, even after 22 years of discussion, the Brazilian Telecommunications Code does not yet satisfactorily respond to the necessities of the sector. With the application of its articles, it has become clear that the code contains many imperfections, which, with the rapid development of radio and television, demand urgent reformulations.

The national business sector, broadcasting associations, the Ministry of Communications and many others in some way connected to the sector have already presented a proposal for a new Brazilian Telecommunications Code, which will soon be submitted to the consideration of the national congress. Although two years have now gone by, there is no guarantee when it will be studied by the congress.

THE PROCESS OF OPENING A RADIO OR TELEVISION STATION

A specific department of the Ministry of Communications has the responsibility of drawing up, supervising and modifying the Basic Plan of Channel Distribution (See Tables 9-1 and 9-2) in medium-wave, tropical-wave, frequency-modulation and television broadcasting.

Since there is little interest in tropical-wave broadcasting, requests for opening stations must be awaited.

However, in the cases of medium-wave, FM and television the situation is the exact opposite. Since there are innumerable requests from all parts of Brazil, the ministry publishes a monthly announcement in the official record, in which those interested in taking part in a tender for the opening of new stations are convoked.

Petitioners have a specific period of time during which they must present the necessary documentation to the regional divisions of the National Telecommunications Department, located in a number of capital cities. The local juridical and technical division then analyzes these requests and sends the processes on to the central broadcasting division in Brasilia. The requests are then examined by this division and presented to the minister for a final decision.

The granting of licenses depends on the decision of the minister, who must take the following criteria, either wholly or individually, into consideration:

Table 9-1. Number of Cities in Each State With Medium-Wave Stations Before and After Implementation of the Basic Plan of Channel Distribution.

State	Before Plan	After Plan (including tenders)
RONDONIA	2	7(5 deles sao mu cipios
ACRE	1	7
AMAZONAS	5	39
RORAIMA	1	2
AMAPA	1	3
PARA	3	46
MARANHAO	1	34
PIAUI	3	25
CEARA	11	36
RIO GRANDE DO NORTE.	4	15
PARAIBA	4	13
PERNAMBUCO	12	25
ALA OAS.	5	9
SERGIPE	2	9
BAHIA	14	37
MINAS GERAIS	81	102
ESPIRITO SANTO	7	21
RIO DO JANEIRO	26	31
SAO PAULO	136	145
PARANA	58	86
SANTA CATARINA	42	57
RIO GRANDE DO SUL	86	106
MATO GROSSO	9	39
GOIAS	13	47
DISTRITO FEDERAL	1	1
TOTAL	**528**	**943**

**Table 9-2. Number of Medium-Wave Stations Allocated to
Each State Under the Basic Plan According to the Unit of the Federation.**

State	Stations Existing Before Plan	Tenders	New Stations	Total
RONDONIA	2	-	8	10
ACRE	2	-	10	12
AMAZONAS	7	-	47	54
RORAIMA	1	-	1	2
AMAPA	2	-	3	5
PARA	7	-	47	54
MARANHAO	5	-	37	42
PIAUI	5	-	26	31
CEARA	21	1	28	50
RIO GRANDE DO NORTE	10	-	12	22
PARAIBA	9	1	10	20
PERNAMBUCO	20	1	13	34
ALAGOAS	8	-	-	-
SERGIPE	6	-	7	13
BAHIA	27	-	27	54
MINAS GERAIS	108	2	24	137
ESPIRITO SANTO	8	-	15	23
RIO DE JANEIRO	50	-	6	56
SAO PAULO	197	3	16	216
PARANA	92	-	43	135
SANTA CATARINA	61	4	13	78
RIO GRANDE DO SUL	123	6	22	151
MATO GROSSO	16	-	38	54
GOIAS	24	-	35	59
DISTRITO FEDERAL	5	-	2	7
TOTAL	**816**	**18**	**495**	**1329**

- The proposed programming of the station
- The statement of financial resources
- The installation and utilization of nationally-produced equipment
- The fact that the members and directors of the society truly live in the locality in which the station is to open for a period of at least two years and that the majority share of the stocks or quotas of the outstanding capital of these societies pertains to these individuals

These criteria are published in the original announcement and are known to all the participants.

The minister of communications has the power to grant licenses to FM and medium-wave stations with power capacities of between 250 and 500 watts. Those of more than 1 kW, whether FM, MW, TW, or SW, receive licenses only with the approval of the president of the republic.

ON THE RENEWAL OF LICENSES

Recent years have witnessed a true revolution in Brazilian broadcasting. This movement got underway when the validity of radio station licenses was extended to 1972. Law No. 5,785 (June 23, 1972), which was detailed by Decree No. 71,136 (September 23, 1972), dealt with the subject in more specific terms. At that time, a total of approximately 1000 stations were to have their licenses renewed. Although it was impossible for the ministry to carry out a detailed analysis of each of them, it did provide the government with the opportunity to study the technical and juridical situation of these stations.

In order to facilitate and organize the task, the stations were divided, according to their power capacities, into three specific groups:

● All tropical-wave and medium-wave stations of more than 10 kW (until May 1, 1973);

● All shortwave and medium-wave stations with power capacities between 1 and 10 kW (until November 1, 1973);

● All FM and medium-wave stations of 100, 250 and 500 watts (until May 1, 1974).

In order to obey this schedule, the ministry set up a work group formed of 15 lawyers supervised by an expert in broadcasting who labored at the task on a full-time basis. At that moment, the exhaustive task, which was to last for five years, of normalizing the situation of all the nation's broadcasting stations began.

The problem was less serious in the case of television stations since, in comparison to the approximately 1000 radio stations, there were only 28. Both groups possessed the same juridical and technical problems. The situation of the television stations was normalized during the course of 1977 and 1978.

At the present time, the number of stations operating in each service in Brazil is:

981	medium wave
35	shortwave
105	tropical wave
248	FM
116	television

Of this total, 1454 are in private hands, including the state and municipal governments, while 31 belong to the federal government.

The principal television networks having national coverage are shown on the maps in the illustrations accompanying this chapter.

NATIONAL INTEGRATION

In 1975, by means of law No. 6,301, the government created the Brazilian Broadcasting Company, *Radiobras*, with the following objectives in mind:

● set up, operate and explore the broadcasting services of the federal government

● set up and operate its own broadcasting relay and retransmission networks, exploring the respective services

● broadcast educational programs produced by the federal government, while also producing and broadcasting informative and recreational programs

● encourage and stimulate the training of specialized personnel demanded by broadcasting activities

- carry out other similar activities attributed to it by the Ministry of Communications
- provide specialized services in the area of radio broadcasting.

According to the aforementioned law, the Radiobras stations should operate with the highest possible technical standards, providing radio coverage to areas of low population density and reduced commercial interest as well as to those areas considered to be essential to the policy of national integration. In other words, priority is to be given to the establishment of radio stations in the entire Amazon region.

Once Radiobras had begun operations, the government determined that its priority ojectives be immediately attained—the coverage of those areas of the Amazon Region which lacked radio broadcasting. Directive No. 1,287/77 of the minister of communications authorized the setting up and operation of 10 medium-wave stations between 10 and 50 kW, together with six tropical-wave stations, all of which were to be located in the Amazon Region.

The latter stations also have the same purpose of providing coverage along the nation's borders. Within a short time, these measures will have the result of making it possible for the population of the region to listen to national radio broadcasting, whereas, up to now, the only stations which could be picked up in the area were those of foreign origin.

According to one of Brazil's leading authorities in the field of radio broadcasting, Professor Saint-Clair da Silva Lopes, the State of Amazonas still possesses one of the most primitive forms of communication. In the same context, he mentions the words of a television journalist who stated that the isolation of the city of Manaus (the capital of the state) is the principal, if not the only, factor responsible for the backwardness of the newspapers and radio stations in Amazonas. Almost without communications with the rest of Brazil, Manaus is connected to the rest of the world and thus receives its information from outside the country, while receiving very little from the rest of the nation.

Outside information does not come from nearby, but rather from Washington, Havana, Peking, London and Moscow. The vehicles used in this flow of information are The Voice of America, Radio Havana, Radio Peking, BBC, and Radio Moscow. These stations possess thousands of listeners in Manaus and among them are to be found the newspaper and radio editors of the capital city.

Since these stations transmit programs specifically designed for the region, the broadcasts constitute a true cultural invasion. It is hoped that Radiobras will carry out its mission in such a coherent and convincing fasion that, once and for all, Amazonia will be integrated into the reality of the rest of the nation.

RADIO AND TELEVISION PROGRAMMING

One of the greatest concerns and principal goals of the Ministry of Communications is an improvement in the quality level of radio and television programming. Conscious of the immense influence which these two media have on the formation of the nation's youth, the government has sought to do all in its power to raise program quality.

According to law, the objective of Brazilian broadcasting services is educational and cultural in nature. Thus, even when presenting informative or recreational programming, these services are considered to be in the national interest. As a result, the commercial utilization these services is permitted only to the extent that it does not hinder the fulfillment of the national interest and the legally stated objectives.

At the same time, the specific limits set down as regards obligatory educational programs (minimum of five hours weekly) and informative programs (5% of total daily programming) must be strictly observed.

The legislation also demands that commercial publicity may not occupy more than 25% of the daily programming, or 15 minutes per hour. In order to avoid abuses and the saturation of the audience, each commercial break may not extend for more than three minutes.

The stations are also obligated to enter, free of charge, into the National Radio Network, when ordered to do so by the competent authority. At the same time, they must keep files with the recordings and texts of their programs for periods stipulated in the law.

Fig. 9-1. Cities with Globo TV Network generating stations.

OBLIGATORY GOVERNMENT PROGRAMS

In 1930, the revolutionary government of President Getulio Vargas instituted a radio program to be beamed to the entire population, which dealt with the accomplishments of his government. This program, *The Voice of Brazil*, still exists today and goes on the air on a daily basis with the exception of Satrudays, Sundays and holidays.

It consists of three distinct parts, each of which is dedicated to one of the three branches of government. The executive occupies the largest share (30 minutes), with the remaining 30 minutes being divided equally between the legislative and judiciary branches. The program presents the principal events of the day, together with the resolutions taken by the government and any other events which deserve mention.

Parallel to this program, there is another which is educational in nature and has the finality of teaching basic literacy skills to adults. The preparation and presentation of this program is the responsibility of the National Tele-education Program (PRONTEL), which operates under the auspices of the Ministry of Education and Culture.

Every radio station in the country is obligated to enter the national network and broadcast these programs, under pain of possible punishment.

THE SOCIAL IMPACT OF RADIO BROADCASTING

At the beginning of this chapter, we mentioned the territorial extension of Brazil, the three obscure centuries of colonialism, and the difficulties in the path of expansion in the field of communication.

Obviously, a nation of the territorial dimensions of Brazil faces an almost endless series of difficulties which hinder the attainment of progress and development. The first challenge which must still be overcome is that of the size of the country. The first towns and cities arose as a consequence of the arrival of the Portuguese colonists and the later foreign invasions. As a result, the largest population centers to be found in Brazil today are located along the coast. Another cause of this is the fact that there exists a large barrier of mountain chains separating the coast from the interior.

Many years passed before the federal government was able to more effectively improve and integrate the interior regions into the life of the nation's large urban center. Highways and railways were built for the purpose of joining the less developed areas of the country to those which enjoyed a greater degree of prosperity. However, even today Brazil suffers from insufficient rail transport and the inefficient utilization of that which already exists. Of course, this is of prime importance, for the rail system not only shortens the distances between the dimerse regions and facilitates the out-flow of production, but also makes possible closer communication and greater integration among the different areas of the nation.

It should be emphasized that it is through telecommunications that the Brazilians have really come to know one another. In the beginning, a few radio stations installed in the interior were the only real sources of news and information for the local population. Later, such phenomena as the telephone, cable and microwave systems introduced fantastic improvements in the telecommunications structure as a whole.

The first television station appeared in Rio de Janeiro in 1950; others were soon opened in Sao Paulo Belo Horizonte, and other cities. However, it was only in the 1960s, and more specifically, after 1965, that this means of communications began to grow rapidly. Cities with Globo TV Network generating stations appear on the map in Fig. 9-1, and cities with Globo Network stations are shown on the map in Fig. 9-2. The maps in Figs. 9-3 and 9-4 show cities in the Tupi Network.

The high technical quality of the Brazilian television networks had the immediate result of encouraging sales of sets among all the Brazilian economic classes.

Traveling about the nation, it can be seen that in the large cities and even in the most distant villages, the horizon is dotted with innumerable television antennas, a sure sign that informative and recreational programming is being received in all parts of the country's vast territory.

Fig. 9-2. Cities with stations that transmit Globo TV Network programming.

Fig. 9-3. Cities with Tupi TV Network generating stations.

The Brazilians were able to watch as men first stepped onto the surface of the moon, witness the play-by-play of the World Cup Matches in Germany, Mexico and Argentina, as well as follow the daily news events from the four corners of the globe, for like any other people, the nation's population anxiously accompanies world events.

Television and radio have been able to unite the different regions of the nation, and the country as a whole, with the rest of the world. The only area still lacking these services is the Amazon Basin, but this area will soon be incorporated into the system through the efforts of Radiobras.

As an important mass media, broadcasting has demanded that the government take special care in seeing to it that those responsible for these services are well aware of the fact that they have enormous influence over the actions and thought of the Brazilian people.

In the words of one of the pioneers of Brazilian broadcasting, Roquete Pinto: "Radio (and now television) is the newspaper of those unable to read; the teacher of those unable to attend school; the free recreation of the poor; the source of new hopes; the consolation of the ill; the guide of the healthy; and thus it must always be carried out in a spirit of high quality totally devoid of egotistical interests."

One of the principal concerns of the government is to make those responsible for broadcasting services conscious of the role they play in the formation of the country's youth. In 1975, upon addressing the students at a school of communications, the minister of communications expressed himself in the following way: "The policy adopted by Brazil in the field of broadcasting is based on the social responsibility of the private sector, in which the state, while safeguarding the right of expression, sets down the principles which guarantee the social utilization of the media, thus making them responsible both for the content and the consequences of the program transmitted."

Those carrying out these services are responsible for the programming and for the value of the message being transmitted. Obviously, they will receive the praise that is due them for that which stands in defense of the common interest and the good of the community. However, in the same way, they are responsible for the human and social problems which their services may cause in the community, and for their lack of participation in the process of development and the improvement of those who make up their audience.

Fig. 9-4. Cities with stations that transmit Tupi Network programming.

Freedom of expression is not equal to, nor may it be transformed into, total immunity, to be used for personal advantage, the benefit of foreign groups, the sale of unsuitable products or the causing of conflict where it has no place. The station is responsible for the transmission of damaging subliminal ideas, false sentiments or unacceptable dissension.

THE FUTURE OF BROADCASTING

The organization, structuring and expansion of Brazilian telecommunications took place in the 1960s. In the first place, the Brazilian telecommunications code was issued, followed by the Federal Administrative Reform, which created the Ministry of Communications, the organ responsible for the carrying out of the sector's policy.

The precarity and stagnation of broadcasting organizations was not only of concern to the businessmen involved, but also led the government to carry out a substantial reform. The principal areas affected were the more rational utilization of channels, particularly on the medium-wave band, opposition to the indiscriminate use and approval of the utilization of frequencies and power capacities which had the result of causing interference in other already existing station, the necessity of decongesting airwave traffic in some regions, the increase of border stations as a means of limiting the foreign cultural invasion, the growth of the national broadcasting industry and, finally, the issuing of laws truly in keeping with the present state of development of Brazilian broadcasting services.

Some victories have already been attained, such as the signing of the Basic Plan of Channel Distribution for Medium-Wave stations which, to a great extent, is already in operation. By means of this plan, it is possible to redistribute frequencies in such a way that they are not only being adequately used but there are also frequencies available for future utilization. Obeying a predetermine schedule, the power capacity of the different stations is also being increased.

With regard to the revision of existent laws, the project has already been completed by the Ministry of Communications, and should soon be sent to the national congress for debate and voting.

The tables accompanying this chapter show the distribution of channels on the medium waves before and after the new plan went into effect.

In the near future, the ministry will also have a radio monitoring headquarters, which will include stations at eight different sites in Brazil. These stations are already being set up. The network has the purpose of controlling radio communications, detecting interference, observing and picking up clandestine transmissions which could harm the transmission of the legally established stations or constitute a threat to national security.

According to statistical data, last year there were 39,200,000 radio receivers and 60,000,000 television viewers using a total of 10,800,000 black-and-white sets and another 1,600,000 color sets in all of Brazil. The color system used is the PAL-M, developed in Germany.

During recent years, Brazil has managed to gain a position of equality with the more advanced nations of the world in terms of telecommunications.

196

Chapter 10
Broadcasting In Guyana

Ron Sanders
Communications Consultant
Government of Artigua

Ron Sanders was program director of the Guyana Broadcasting Service from 1971 to 1973 and general manager from 1973 to 1976. He was also president of the Caribbean Broadcasting Union, 1975-76; and a director of the Caribbean News Agency, 1976-1977. He lectured in communications at the University of Guyana in 1976 and was public communications officer at the Caribbean Development Bank in 1977. He is now communications consultant to the Government of Antigua. Mr. Sanders is also author of *Broadcasting in Guyana* published in July, 1978, by Routledge — Keegan Paul, London.

By the time this is published, the Government of Guyana will own and control broadcasting with all the censorship, program direction and editorial management that ownership and control connote.

At the time of writing, the government owns and controls one of the two radio stations, the Guyana Broadcasting Service (GBS), and on January 1, 1979, it nationalized the other radio station, Radio Demerara, which has been owned for over 25 years by a British company, Rediffusion Ltd. There is no television in Guyana.

Over the last four years, the Guyana Government, which embarked on a program of transforming the country from a free-enterprise system to socialism, has taken increased control of the mass media. Recently, the scale of control has been widened and it now appears that the government will dominate all the news media.

ENVIRONMENT FOR BROADCASTING

Guyana is an 83,000-square-foot landmass on the north coast of South America. It is sandwiched by Venezuela, Brazil, Surinam and the Atlantic Ocean. It is also the only English-speaking country in South America. Its population is roughly three-quarters of a million (accurate figures are not available, since the last census was done in 1970), the majority of whom are under the age of 30. Only the coastal strip is of any social or economic importance at present, for on it lives 90% of the population. (Peter Newman, *"British Guiana—Problems of Cohesion in an Immigrant Society,"* Oxford University Press, 1964.)

The country was colonized by the Dutch and then the British, who ruled it for 350 years. It was colonized to produce sugar and to this day, sugar continues to be one of its two biggest exports. The other is bauxite.

The country's interior is known to be rich in minerals and timber. The land is fertile and has often been described as a potential "bread basket of the Caribbean." However, there are no roads to the interior and no facilities such as running water and electricity. Consequently, the Guyanese people have shown no interest in venturing into the interior to develop it.

The main races of the country are Indians, decendants of East Indians imported from India by the British as indentured labor for the sugar estates after slavery was abolished, and Africans, the decendants of the African slaves. These two races have co-existed with a great deal of hostility which flared up in racial violence between the years 1962-64. The two major political parties, the Peoples National Congress (PNC) and the Peoples Progressive Party (PPP), raise their support on racial lines.

The country became independent of Britain on May 26, 1966, and in February, 1970, it declared itself a republic. Its system of government has a "ceremonial" president, elected every five years by a majority vote in the unicameral parliment. Members of parliment (whose terms are five years) are returned at general elections across the country under a system of proportional representation, based on adult suffrage. There have been two general elections and a referendum since independence, all of which the opposition party claims were rigged by the ruling party.

Over 80% of the economy is controlled by the government, which in pursuance of its socialist policy has nationalized all the major productive sectors of the conomy. The government has also established its own bank and insurance company, and there have been recent statements that the foreign-owned banks and insurance companies will shortly be nationalized. Because of its nationalization policy, and also because the government instituted a program of import substitution (i.e. not importing any products which could be substituted locally), competition among goods and services in Guyana practically disappeared in the last five years.

In 1964, a UNESCO team did a survey of broadcasting in Guyana and estimated that $1.5 million (Guyana) existed for advertising. Of that sum, 50% was allocated to radio. However, only 30% originated from local advertisers, the other 70% coming from foreign sources. (UNESCO Report, *British Guiana Broadcasting*, 1965.) By 1976, both radio stations, which had hitherto been returning profits, were showing enormous losses, due to the reduction in the number of goods and services available for advertising.

Prior to 1968, all the mass media in Guyana were privately owned. The government then took away from Rediffusion Limited the license to operate one of the two radio stations which it had at the time. The Guyana Broadcasting Service was then used as a government station.

In 1971, the government bought one of the two national daily newspapers from a local businessman and *The Chronicle* became a government newspaper. In 1974, the government moved again to nationalize *The Guyana Graphic*, owned at the time by Thompson Newspapers of Canada. The two newspapers were merged into one. The government, it should be noted, did not provide financing for any of the mass media which it nationalized; it simply instructed the boards of the companies to operate under the same commercial lines as before.

EVOLUTION OF BROADCASTING

Only four years after the British Broadcasting Company (as the BBC then was) started operations in 1922, experimental broadcasting was going on in Georgetown, British Guiana (as Guyana then was) for two hours a week and programs were relayed from Daventry in England with fair results. (*Handbook on Broadcasting in the Colonies*, 1958, Information Department Colonial Office, London.)

After the Second World War in 1948, the British Government recognized the need to build new broadcasting stations in the colonies "because false rumors and impressions were believed in the remotest parts of many territories and there was no way of learning the truth." A medium-wave transmitter was then brought into service.

But, broadcasting did not start on a regular and daily basis until January 1, 1950, when a 15-year franchise was granted by the British Government to the British Guiana United Broadcasting Company Limited, a subsidiary of Rediffusion Limited of London, to own and operate a private commercial radio station. The agreement was tailored to meet the needs of the British Government and included, for instance, a provision making it compulsory for the station to broadcast not less than 21 hours per week of the BBC's programs. It also demanded 10½ hours of air time per month for the government.

The license arrangement to the subsidiary of Rediffusion was far from satisfactory, as it made no reference to program standards in the public interest, except the usual one of defamation and obscenity. In 1958, the company introduced a second radio station. This one was called the British Guiana Broadcasting Service (BGBS) and the government at the time merely negotiated for 10% free broadcasting time on the station. The second radio station was introduced to accommodate the large number of commercials which the one station could not handle.

By 1964, sufficient concern was expressed about the role of radio stations in a developing society to encourage the government to seek assistance from UNESCO in determining a course of action in regard to broadcasting. The two stations at the time were largely disc-jockey operations featuring hours of music, interspersed with commercials and cheap soap operas. A UNESCO survey was taken in 1964 and a report, *Broadcasting in British Guiana*, was submitted in 1965. The terms of reference of the UNESCO mission involved a request for advice on the following:

> The reorganization of local sound broadcasting along public service lines and especially on the early replacement of some overseas programs by locally produced programs using local talent. (UNESCO Report, "British Guiana Broadcasting," 1965.)

The report of the UNESCO team of Andre Quimet and P. Lloyd Grant noted that "the government's objectives regarding sound broadcasting appear to stem from a dissatisfaction with the present radio program service as expressed in a government letter advising the private broadcasting company that its franchise would not be renewed after December 31, 1964." In fact, the franchise was renewed for a further four years, until 1968.

The UNESCO report went on to note "the government felt that while this company had done valuable work over the years, especially in the development of technical installations which provided a reasonably satisfactory countrywide service, its program policy has been less than satisfactory. It was felt that in a strongly commercialized program pattern, adult education and cultural broadcasting has been neglected."

The UNESCO team's suggestions to the Guyana Government were not accepted. The suggestions were:

1. A government monopoly financed partly from advertising revenue and partly from public funds and license fees
2. A shared system with transmitting facilities owned and operated by private interests but programmed partly with programs financed and produced in government studio facilities, and partly with programs produced by the same private interests and financed from commercial revenue.

Rediffusion continued to own and operate the two radio stations until 1968, when the government took over the operation of the Guyana Broadcasting Service.

The 1964 UNESCO team had also been asked to consider the possibility of television being introduced; they were clearly not in favor of this idea. However, since recommendations on television were within their terms of reference, they suggested that "the time at which television should be introduced be decided on the basis of the relative priority attached to it by government in its list of social and economic development projects and of its expected contribution to such development."

Three years later, two local experts, Hugh Cholomondeley and Kit Nascimento, submitted a report to the Guyana Government recommending the establishment of a national broadcasting system to include both radio and television. The government considered this proposal for a while and even sent a number of people for training in television production at the BBC in London. The argument of the two local experts was that radio revenues would pay the cost of television, and government needed only to provide the capital costs of establishing studios and equipment. One year later, however, in 1968, the government did not consider television a priority and the idea of a national broadcasting service of radio and television was dropped.

200

When the government finally got its own radio station—GBS—started on October 1, 1968, the sum of $80,000 (Guyana) was loaned to the operation. No other funds were paid by the Government. The station began public service broadcasting along the lines the government wished to see. However, the cost was being paid by commercial advertising. The accent was placed on local programming, and local talent in drama, music and comedy was utilized. This continued until 1972, when, because of the government's policy of import restrictions, advertising revenue began to dwindle. By this time, the government had nationalized the two privately owned newspapers and there was increasing censorship of the media. The government-owned station was particularly subject to governmental control, as a minister of the government was the chairman of its board of directors, and the minister of information retained control over policy.

In 1973, the Guyana Government had become the focal point of attack for censorship of the press. The Caribbean Publishers and Broadcasters Association (CPBA) commissioned a report on press freedom in Guyana from Dr. Everold Hosien, who was then a lecturer in communications at the University of the West Indies in Jamaica.

Dr. Hosien's report was published in 1975. In it he claimed to have found several types of control being exercised on the media in Guyana by the government. Among the controls were:

● **Covert control:** The evidence indicates that media personnel, on the basis of past experience, have become aware of what the owners (i.e. government) would or would not like to see in the news. These respondents know that government is sensitive to dissenting opinion which criticizes government activities. As a result, no (or limited) coverage is given to dissenting views. Media personnel are job conscious and, therefore, choose not to disturb their "tenuous" positions. We conclude government has exercized covert control over the expression of dissenting opinion in the press.

● **Overt control:** The evidence indicates that media personnel have been specifically directed by a government minister on what not to cover in the news. Many of these directives deal with not covering public meetings or statements in which dissenting views have been expressed. These directives often have been issued via the telephone, or more subtly, at cocktail parties. (Hosien, Everold N., *The Implications of Expanded Government ownership of the Mass Media for Freedom of the Press in Guyana*, January 22, 1975.)

There is no legislation to deal specifically with broadcasting or to provide an adequate means of regulating the use of the airwaves in so far as program content and program standards are concerned. The law relates to broadcasting only indirectly under the general provision of freedom of expression; it also covers broadcasting under general provisions for defamation. A broadcasting act was prepared by a local media expert in 1969, but it was not accepted.

In 1975, the author of this chapter, who was then the general manager of the Guyana Broadcasting Service, recommended to the government that radio needed to be regionalized. It had already been pointed out that the population of Guyana was subdivided into various groups: rural agricultural workers, who were also primarily of East Indian origin; urban middle class in the capital city, Georgetown, and its environs, who were mostly of African extraction; a mining community in the Linden area who had grown up apart from the rest of the country; the Amerindians in the deep interior of the country. The proposal had been that in order to communicate effectively with each group, programs had to be tailor-made for each community. Therefore, it was suggested that small radio stations be established in each community capable of feeding program material into each of the other groups and also with the facility for carrying joint national broadcasts.

The government agreed to fund the capital costs for establishing the regional radio stations, and a start was made in Linden. Unfortunately, the Guyana economy took a turn for the worse and the station was not completed. However, since 1976, the mining community at Linden has benefitted from a 10 kW medium-wave transmitter which relays program material from the Guyana Broadcasting Service with a very good signal. Previously, the people of Linden could not hear either of the two radio stations in Guyana and instead listened to programs from Venezuela and Brazil. The communities in the interior of the country still receive only Venezuelan and Brazilian stations and their knowledge of events in Guyana is usually days old and based on word-of-mouth presentations from visitors.

BROADCASTING TODAY

The present system of broadcasting comprises the two radio stations (Radio Demerara and GBS) both of which are in competition with each other for listeners and revenue. While there is no broadcast policy laid down to govern either radio station, in practice the minister of information is responsible for broadcasting and can give directions to the government station, GBS. Under Radio Demerara's licensing agreement, the minister can, in theory, order that certain material be broadcast or not broadcast, and since the minister has the power to revoke the license under which Radio Demerara operates, the station has been responisbe to ministerial direction.

Radio Demerara's policy has been guided by its shareholders' demands that it makes a profit. Consequently, it has programmed cheap and popular programs. However, over the years since Guyana became independent, it has moved in the direction of more locally produced programs to protect its license. Caught, as it was, in the dilemma of making profits and serving local needs, it did very well until recently when the economic situation turned bad and advertising dwindled as a result of government policy.

The station has no written policy, except that it should make profits. From time to time, the board of directors makes policy decisions on matters of programming, but largely this is left to the general manager. The station suffered the additional disadvantage of an uncertain future. Its operators knew that it was only a matter of time before the station would be nationalized. The uncertainty of the future dictated that the station could not commit itself to an expansion of its facilities or a long-term program policy.

The Guyana Broadcasting Service has been guided in matters of programming by its general manager with a board of directors responsible for the financial performance of the station. However, the minister of information is officially responsible for policy and directs the station in this regard. The policy decisions of the minister of information have included directing that certain news stories not be carried, nor programs which featured criticisms of the government and even records.

When the station began operations in 1968, in a healthy advertising climate, it geared itself to serving the public needs and financed its work from local advertising. The station introduced local drama on a regular basis, and pioneered public participation programs as well as investigatory reporting and documentaries. However, as the revenue position became worse, the station depended on the government to take up the shortfall in its income. That dependence on the government encouraged even greater control by the ministry of information over matters of programming.

The two stations offer basically the same coverage to the people of Guyana. The transmission coverage on medium wave gives good quality reception on the coastal regions only. The stations also offer a shortwave service to the interior, but since it is not the habit of the inhabitants of the interior of the country to tune to the shortwave channels, this service is virtually ineffective. Both Radio Demarara and GBS also offer FM services to Georgetown and its immediate environs, but the same program feed is carried on medium wave, shortwave, and FM.

No survey of listeners to the radio stations has ever been made, and it is impossible to give an accurate figure about which is more popular and for what reasons. It is also difficult to state an accurate figure of the number of radio sets in the country, because the system of collecting license fees for radio sets has always been grossly inefficient. Since 1960, the post office, which does the collection, has been giving the figures at 100,000. However, it was widely felt that every Guyanese family has at least one

Table 10-1. Comparative Analysis of Radio Demerara and BGBS.

	RADIO DEMERARA (in hours)	BGBS (in hours)
BBC news and news analysis	4 2/3	3
Other BBC relays	1½	-
BBC transcriptions	7	1
Other transcriptions (Canada, USA, Australia), mostly soap operas	20½	8
Local live studio productions mostly news and live sport	14 1/3	7
Local gramaphone record programs; mostly foreign records	64	44

radio set. Cheap transistors also encourage the belief that most families have more than one radio set. In fact, the advertising rate card of the Radio Demerara lists the number of radio sets in the country at 375,000.

In 1964, when the government invited a UNESCO team to study the role of broadcasting in Guyana, it had indicated that it wished to see programming in the public interest, including adult education and cultural programs. However, the UNESCO team summed up listeners's tastes and habits as follows: "Except for a small minority, radio listeners everywhere are more interested in being entertained than they are in being educated or even informed through the radio." An analysis of the weekly program schedule of Radio Demerara and BGBS (as GBS then was) in 1964 is revealed in Table 10-1. The 1977 analysis of the weekly program schedule of Radio Demerara and GBS in Table 10-2 reveals the breakdown of programming between the private and government stations using the UNESCO format of 1964. (Sanders, Ron, *Broadcasting in Guyana*.)

From the two analyses, it can be seen that while there has been an overall increase in local programs, the number of cheap and popular programs also increased. In the case of the government-owned GBS, its program format is more oriented to the local situation than that of the foreign-owned Radio Demerara. However, neither of the two stations achieved the lofty ideals of programming which the government, in 1964, had indicated it wished to see.

There is one feature of the effect that commercial support for radio has on programming which is worthy of particular attention, given Guyana's history of racial tension. Both radio stations have allocated specific broadcast times for East Indian programming. By this, they mean that at specific times only music originating from India (usually connected with East Indian films), is played on the air. The program slots for East Indian music have always been regarded by the East Indians as "our" time and by the Africans, particularly as "their" time.

In a paper presented at a seminar in Guyana, sponsored by the International Institute of Communications, the author has argued, "It would seem that an effort to build a nation from diverse races, particularly in a situation where the two major races have a history of conflict, that the radio stations would have attempted to bring about some degree of empathy between the races by exposing them to each other's culture. It would have been a simple matter to disband the Indian programs and include the music in the broad format of the station's programming. Instead, what exists is a system which emphasized the racial differences." (*Towards a Communications Policy*, paper, IBI Seminar, Guyana, Ron Sanders, 1974.)

This view was reinforced by a study conducted by communications students at the University of the West Indies and reported by Dr. Everold Hosien in an article in the Commonwealth Broadcasting Journal, *Combroad*. In his report, Dr. Hosien says, "In the first instance separate ethnic programs reinforce the uniqueness of Indian culture and give it legitimacy. Separate broadcasts, however, may not lead to national acceptance of Indian culture as part of national life, to be understood and respected. A

Table 10-2. Program Analysis of Radio Demerara and the Government-Owned GBS.

	RADIO DEMERARA (in hours)	GBS (in hours)
BBC news and news analysis	2 ⅓	-
Other BBC relays	-	-
BBC transcriptions	½	-
Local analysis (news)	¼	1
Other transcriptions (Canada, USA, Australia), mostly soap operas	6	-
Local live studio productions mostly news and sport	10 ⅓	17½
Local gramaphone record programs now mostly Guyanese and West Indian	53	61
Local dramatic productions (plays, short stories)	1 ¼	2½

recent survey confirms conventional wisdom that it is primarily East Indians who listen to Indian programs. Other ethnic groups simply ignore these programs and may even resent their presence."(*Combroad*, April-July, 1976, page 12.)

In 1976, GBS attempted to disband its East Indian programs ad to incorporate East Indian music within its broad format. However, the mainly East Indian advertisers who paid for these programs, withdrew their advertising support from the station entirely and the station was thus forced to reinstitute the programs, as it could not find the revenue elsewhere.

Financially, both radio stations have been doing very badly. In 1976 and 1977, both stations returned heavy losses. The financial performance had a disastrous effect on programming and there was a severe cut-back in all costly programs including local drama, documentaries, and coverage of sport.

FUTURE OF BROADCASTING

The Government of Guyana has now nationalized Radio Demerara, the former Rediffusion-owned station. In strictly economic terms, the move makes a great deal of sense. Rediffusion clearly could not continue to run a radio station at a loss and, as the two stations were sharing what little advertising revenue is left, neither was really benefitting from continued competition. If one station gets the complete advertising revenue available, then it might be able to survive.

However, in terms of being free to report events without bias and also to criticize government actions, the nationalization of Radio Demerara is a sad blow for broadcasting. With total ownership and control and no mechanisms established by law or otherwise for broadcasting to operate independent of the dictates of the ministry of information, it is clear that government censorship will increase.

There will be no expansion of the broadcasting facilities for a number of years, as the Guyana Government is in dire economic straits. At the end of 1977, the country's balance of payments continued to be very unfavorable with a deficit of U.S. $90 million. There was an overseas debt of U.S. $270 million, plus much more unpaid commercial debts and a government deficit of U.S. $30 million. Clearly, there is no money to be spent on the development of broadcasting.

Therefore, broadcasting will now be a tool of the government; its purpose will be to brighten the image of the party in office and to deny access to dissenting views. In these circumstances, broadcasting will play a negative role in Guyana's development.

Chapter 11
Broadcasting in India

Mehra Masani
Vice-President
International Institute of Communications

Most of the problems of Indian broadcasting can be traced back to the nature of its origin and the circumstances in which it developed. For this reason, it is necessary to go back, briefly, to the early years of radio in India.

The year, 1927. India, a colony of the British Empire, was governed efficiently, if rather unimaginatively. A few Indian businessmen in Bombay and Calcutta decided to introduce a very limited radio service, but soon had to abandon it because, in a coummunity with very low purchasing power, the number of radio receivers was so limited as to make service unviable.

Meanwhile, in some parts of the world, radio was being used for propaganda of one kind or another and the Government of India thought that it might be a useful tool for publicizing its views and activities to counter the growing demand for independence.

DEVELOPMENT OF BROADCASTING

The Indian State Broadcasting Service was established in 1930. Considering the huge territory to be covered (Pakistan and Bangladesh were also part of Britain's Indian Empire at that time) and the enormous number of languages and the diverse cultural patterns, broadcasting could only have been organized and financed by the government.

Commercial radio was out of the question because of the poverty of the country, nor could an autonomous broadcasting organization like the BBC function in the political conditions prevalent in a colony. Thus, radio came to India under government auspices and so it has remained ever since.

In 1934, the government decided to reorganize the service on proper lines and asked the BBC for the services of an experienced broadcaster. In August, 1935, Lionel Fielden of the BBC came to India as the first controller of broadcasting. He was an idealist, fully committed to the idea that broadcasting was a public service to promote the public good, in the broadest sense, and not a tool for government publicity and propaganda.

However, the control of broadcasting was not in the hands of the controller but in those of the British civil servants who governed India at that time. It was not long before Fielden clashed with them and left India in 1940, a disillusioned man. The government considered it a good riddance because, taking the cue from the Nazi experiments with radio propaganda, it had already decided to use radio to project the empire.

During Fielden's regime, when the service was renamed All India Radio (AIR) and for some years thereafter, there was a modest expansion of the radio network. By 1947, the number of stations had increased to 12. There were also five stations in the Indian states which were still ruled by Indian princes. The number of radio sets went up from 38,000 in 1936 to 74,000.

The expansion was accompanied by the introduction by Fielden of several instructional programs for special groups of listeners such as women, children, schools and farmers. Such programs had already been started in other countries with very different conditions. To be successful, they should have been adapted to Indian conditions; but, because there was no well-considered policy for broadcasting, all such programs were started without adequate preparation and sufficient funds.

The expansion of the network was also very slow during this period because the government of India failed to realize the potential of radio. The main thrust during this phase was to use radio in support of the war effort of the Allied powers. A central news organization was set up to prepare bulletins under proper supervision, and several services in foreign languages were started to counter Japanese propaganda in Southeast Asia and German propaganda in West Asia. Propaganda concerning the war was accompanied by the projection of the British stand on Indian independence. None of the nation's leaders had access to the radio.

CHALLENGE LOST

When India became free in 1947, the whole question of how to use the technology of radio communication should have been considered in detail and plans made to change the system to fit the changed conditions. It would be incorrect to say that the system was transferred from Britain, because the BBC model was not transferred to India, even though Fielden tried, in many ways, to introduce BBC standards.

The government of free India, in 1947, was unaware of what it could do with radio in the service of the country. At this time, if the decision had been taken to use the radio to communicate with the people to promote their welfare by increasing their political awareness, improving their work techniques, encouraging social change and progress, the history of broadcasting in India would have been very different. In fact, broadcasting continued with the same structure, staff, program pattern, administrative and budgetary practices as before, as if nothing had changed except that a new set of political personages regarded it as a tool for *their* publicity and propaganda. Thus, the opportunity to adapt and use a powerful technology for Indian needs and goals was lost.

POSTWAR BROADCAST SERVICES

The plans for the development of broadcasting from 1947 onward, which should have embraced the entire nation, were mainly restricted to providing a service to the cities. Every regional language was provided for by installing pilot stations of only 1 kW power in the capital city of each state of the union. To some extent, this satisfied the demand of the educated urban listener for information and entertainment and also provided a number of musicians, talkers, writers, and others to express themselves and earn some money.

By 1950, there were 21 stations broadcasting in all the major languages. Higher power transmitters were installed at some regional and zonal stations to extend their coverage. From 1951 onward, the pilot stations were replaced by high-power transmitters at most stations.

When the first 5-year plan was issued in 1953, broadcasting was not mentioned. The need and the urgency to use radio to communicate with the people, for whose benefit the plan was made, was not even recognized. Only 30% of the country's area was covered by AIR's medium-wave service. However, the second plan, issued in 1956, referred to the plans for the expansion of broadcasting by installing higher power transmitters and increasing the number. But the structure remained the same—a highly centralized broadcasting service in which programs emanated from a few urban centers.

In 1962, All India Radio started an experiment with Radio Rural Forums. The idea was borrowed from Canada, where such forums had functioned effectively for some time. With the collaboration of UNESCO, the experiment was successful and AIR set up about 20,000 such forums all over the country (Fig. 11-1). Community receivers were provided for the farmers to listen to sepcially prepared programs; these were discussed by the listeners and their questions, comments, and reactions were conveyed to the radio station. However, the forums declined in number and in utility, mainly due to organizational difficulties.

In 1959, an experimental TV service was started from Delhi (Fig. 11-2), giving instructional programs of one hour's duration, twice a week for 21 "teleclubs," which were formed by installing a TV set, in a school or community center, on which people living in the neighborhood could view the programs. Broadcasts for schools were also introduced. Some years later, in 1965, the service was extended to provide information and entertainment for Delhi viewers in general and since then several TV centers have been established in Bombay, Calcutta, Madras, and other cities. In spite of the expansion of the network, which went on during this phase, the impact of radio and TV was so limited that it was evident, even to the government of India, that some fresh thinking was necessary.

ATTEMPT TO BREAK GOVERNMENT CONTROL

The Committee on Broadcasting and Information Media was appointed in 1964 to suggest changes in the structure, the control, and the operation of the broadcasting service. Its report, issued in 1966, made numerous suggestions for radical changes which would make the service:

- independent of the direct control of government by the establishment of two autonomous corporations to run the radio and TV services
- responsive to public opinion by the organization of audience research on a systematic basis
- effective in mobilizing popular support for the plans and projects devised for the economic and social progress of the people by giving priority to programs for rural listeners, including women, rural schools, the under-priviledged in tribal areas, and so forth, by the establishment of local radio stations to serve the specific needs of each district.

The committee's recommendations, made in 1966, were examined by a series of authorities. Some recommendations, such as introducing commercials and the separation of radio and TV, were accepted but the one suggesting autonomy for AIR and TV was rejected for various unconvincing reasons.

With all the expansion which has taken place, even now, only 76% of the country's area, where 90% of the people live, is covered. Of course, that does not mean that 90% of the people listen to AIR. Only those with radio sets or those who have access to the limited number of community sets can do so. In fact, only about 20% of the people have regular access to radio.

Today, there are 84 radio stations, including auxiliary centers, seven TV centers covering about 15% urban viewers. The radio network puts out 1350 hours of programs daily, while several broadcasts in 16 languages go out to several countries in Southeast Asia. Africa, West Asia, Europe and Australia for a total of 23 hours per day.

In absolute terms, these figures are impressive, but in relation to the size and population of India and the crying need for communciation with rural India, the growth of broadcasting is slow and its impact very restricted. Why?

India's experience is not very different from that of other Third World countries. Considering the power of broadcasting to help such countries in their progress to modernization and economic growth, it is essential to consider in some detail in what respects the broadcasting service has been disappointing, particularly in its utility to the mass of poor, illiterate people who live in villages.

Third World broadcasting suffers from state control, which has certain well-recognized characteristics, whether in India or in the Soviet Union or in Tanzania. The objective is both implicitly and explicitly to promote the government's interests as it sees them. Inevitably government control leads to bureaucratic organization, a top-heavy administration, dominated by the civil service, lack of flexibility in planning and production of programs, unimaginative and often crude misuse of the medium for short-term political gains, resulting in loss of credibility, inadequate investment of funds and scant regard for public reactions.

The state control of broadcasting must also result in the recruitment of civil servants to run it, as has occured in almost all developing countries. The dependent relationship of an employee detracts from a sense of responsibility for his actions. The Indian Broadcasting Service, like many

Fig. 11-1. Farmers being interviewed for a rural program broadcast.

others started under government auspices in the colonies of Asia and Africa, displays all these defects. Since the Government of India did not relinquish control even when India became a free country, Indian broadcasting continues to suffer from all these faults.

GOVERNMENT PROGRAMMING

While the disadvantages of a state-run broadcasting service are obvious, they can be compensated for, to some extent, by using the service for education, in the broad sense, which a commercial service could not do. The government, unconcerned with profit and popularity ratings, could cater, through the radio, to that vast mass of illiterate people, living mainly in poverty-ridden villages, unserved by any other medium of information, entertainment, and education. Efforts have certainly been made in that direction, but the results have been uneven and unimpressive.

The government has declared that its program policies have been formulated to project important trends and developments in public life, to provide information and education, to act as a vehicle of artistic and cultural expression, to serve the needs and interests of listeners in the field of entertainment, and to reflect and interpret the national and international policies of the government. These are unexceptionable aims, but whether they are achieved would depend on whether there are enough resources in money, trained manpower and equipment to produce good programs, do systematic audience research to determine listeners' needs and interests, and maintain a flexible program policy to adjust programs to changing tastes and demands.

211

AUDIENCE COMPLEXITIES

Undoubtedly, the problem is vast. Even after the partition of the country in 1947, India has 23 states, some larger than many countries of Europe; it has 16 major languages, some of them spoken by as many as 30 to 50 million people; 51 dialects, and 87 tribal dialects; tribal areas with distinct economic and cultural needs; 80% of the total population in rural areas, at different stages of development and communication with urban areas; castes and classes which cause further divisions even within states and within linguistic groups. Except for the USSR, there is no country with such a diverse population spectrum in language, religion, and stages of development. It was a challenging task to provide a broadcasting service in these conditions and All India Radio has met the challenge with partial success.

For one thing, the service is restricted; it reaches only a part of the population. To own a radio or TV receiver requires a license to be obtained from the post office. The radio license is about $1.50 per year and the TV license is about $5. At present, there are approximately 20 million licensed radio receivers and 500,000 TV receivers in a country with 650 million people. Even if unlicensed receivers are taken into account, there may be about three or four million radio sets. The figures are unimpressive. When they are further broken down between cities and villages, it is found that about 15 million out of the 20 million sets are in the cities, leaving only five million in the villages where over 450 million people live. It means that there is one radio set for every 100 people. The main reason is the high cost of a radio set in relation to rural incomes.

To get over this difficulty, which was recognized even by the British administration, a scheme of community listening was organized by installing a radio receiver at some appropriate place in a village where people could collect of an evening for an hour or so to listen to programs broadcast for them.

But the scheme was plagued by difficulties from the start and has now been written off. At its peak, there were only 90,000 sets scattered over 550,000 villages; the regular and satisfactory maintenance of the sets, particularly the replacement of batteries in unelectrified villages, which constitutes the majority, was beyond the capacity of the state administrations responsible for it; the farmer found it irksome to go, after a hard day's work, to a community center or wherever the set was located; women found it difficult to hear the programs outside their own homes, and so forth. With all the money and effort spent on it, the community listening scheme provided a service to only a small part of the village folk.

Limitations of coverage apart, the utility of the broadcasts is also limited. In most of the cities the radio and TV services are provided on a single channel and all the different types of programs have to be crowded into the 12 hours or so for which the stations are on the air. Time has to be made for programs for women, for children, for schools, for farmers, and other groups, while at the same time, providing news, comment and entertainment (Figs. 11-3 and 11-4) for the general public. Small wonder that everyone feels cheated of his due. The problem is further aggravated

212

by the fact that some of the single-channel stations must serve a variety of linguistic groups which live within their service area.

PROGRAM QUALITY

Because the funds for operating the services are limited by government allocations, the quality of the information and entertainment provided is uneven. Radio and TV audiences complain that the programs are, at times, dull and unimaginative. The most serious charge of all is that because the services are owned and operated by the government, the news and comment is biased and no dissenting opinions are allowed expression whether in political, economic, or even social matters. Unless issues which agitate the educated, urban audience are fully and freely discussed, the broadcasting service cannot play a significant part in public life.

The programs for rural listeners are planned, produced and broadcast from stations located in the cities. Thus, in a state like Uttar Pradesh in North India, with about 85 million people in its villages, scattered over 295,000 square kilometers, there is only one program broadcast from one of the cities in the state. Such a program of about 30 minutes' duration per day can hardly cater for the variety of soils, crops, climate, marketing facilities, and so forth. The program does not, and cannot by its very nature, deal with the specific problems of farmers and artisans in different parts of the state. Being mostly irrelevant to their needs, it is hardly surprising if the program is considered superfluous.

Fig. 11-2. Broadcasting House, New Delhi.

Such highly centralized programs are confined to messages and instruction from the government and its agencies to the people, with scarcely any possibility of a feedback from the people. With limited resources, the broadcasting service cannot undertake audience research to determine whether its programs for the countryside are acceptable and useful.

It is to meet all these deficiencies that local or community broadcasting has been recommended by experts both in India and abroad. The Government of India has, at last, accepted this suggestion and the next few years will see the establishment of such stations. With about 350 such stations, located in the heart of every homogeneous rural area, it should be possible to serve rural India effectively. These low-power transmitters, which are made in India, would carry programs not only for farmers but also for rural women (Fig. 11-5), carrying the message of family planning, nutrition, and so forth. Rural schools, which are inadequately staffed, would also receive appropriate programs at convenient times from the local stations which would serve the entire community. Located within a few miles of each village, the stations could obtain the views and reactions of individual farmers, artisians, women, and young people for broadcast; such stations would also get the much needed guidance for the planning of programs.

Local stations, run by the community, with only a few professional broadcasters to help them, would reflect the needs and interests of the people, thus reversing the present process whereby a few town people plan and produce programs for people with whom they have little contact.

What applies to radio should apply to TV; and, if it is to be used for development, TV programs should also be of local relevance to the communities they serve. To set up hundreds of TV centers is quite beyond the country's resources, and so the government has decided to use satellite for TV communication. It is difficult to see how that complicated and expensive technology with a highly centralized organization can serve small communities in local languages, with programs of local interest, involving local people.

When NASA of the United States offered the use, for one year, of one of its satellites for the Satellite Instructional Television Experiment (SITE) the Government of India used it partly for broadcasting TV programs of an instructional nature for about 5000 villages which were provided with community sets. About half of them had special antennae to pick up programs directly from the satellite.

The evaluation of SITE has not been very encouraging. Apart from organizational problems of maintaining community receivers in good order, there were difficulties in coordination between the various agencies responsible for the hardware and software, and in communicating with people who were unused to the TV medium. There was also a shortage of skilled professional writers and producers to make the programs acceptable. It seemed to most professional broadcasters that a good deal more research and experiment was needed before satellite TV could be considered for prompting rural development.

214

Fig. 11-3. Yakub Ali Khan, classical musician.

However, beginning in 1981, the government has decided that it will use its own satellite (INSAT) for instructional programs for villages. The detailed scheme is not yet available. It is clear, though, that only a limited part of the country can be served in only a few of its many languages. To some professional communicators it appears that INSAT is a mistake. Greater investment in radio would be more rewarding in every respect for a country like India which does not have the infrastructure needed for TV on a nationwide scale, as SITE showed.

As government still controls the broadcasting service, it is not obliged to pay any heed to public opinion. One thing which must be said in favor of the TV policy of the government is that, unlike most developing countries, India does not import any large quantities of entertainment programs from other countries. From the start of television, such imports have been restricted and the greater part of the TV output is indigenuous.

FUTURE PROSPECTS

So much for the past and present. What about the future? 1977 was an important year for Indian broadcasting. The Janata party, which came to power in March that year, included the reorganization of broadcasting in its election manifesto and is thus committed to far-reaching changes and reforms in the system. A working group was appointed in August, 1977, to suggest how autonomy should be granted to both the radio and TV services. It submitted its report to the government in March, 1978.

The report endorses many of the recommendations of the earlier committee and spells out, in greater detail, the nature of the autonomous organization required to provide a satisfactory broadcasting service.

Legislation, based on this report, is still to be introduced in parliament and it remains to be seen to what extent the government is prepared to alter the present system with its antiquated structure and bureaucratic management. More important still is the question whether the reorganized system will be part of a communication system which will make effective use of the electronic media. If the proposals for reorganization are accepted, there will be far-reaching changes in the Indian broadcasting system.

Drawing widely on the experience of other democratic societies, the working group has suggested that broadcasting should be the responsibility of an independent, autonomous National Broadcast Trust (NBT) accountable to parliament to which an annual report and budget would be presented. Like other such organizations, the NBT would be obliged to carry broadcasts required by the government but would have the right to attribute the broadcasts to the government.

Certain legal and constitutional measures have been recommended to safeguard the autonomy of the NBT. It should be declared a corporate citizen of the country so that it can exercise the fundamental right to freedom of speech and expression, which is available only to citizens. The act establishing the trust should be accompanied by a charter setting out its rights and obligations and the autonomy of the trust and its independence of the government should be entrenched in the constitution by an amendment.

A complaints board, to be appointed by the chief justice of India, has been suggested to attend to charges of bias and discrimination in reporting news and commenting on matters of public concern.

The working group was asked to examine the financial implications of autonomy. The dependence on government for funds to run the broadcasting services was always cited as one of the main obstacles to the grant of autonomy. The group has quite rightly rejected this as an argument for continued government control. It gives examples of public services and institutions like the universities, the election commission and the judiciary, all of which enjoy independence from the control of the government, although they subsist on public funds. The group has suggested that all the existing assets of the broadcasting services should be transferred to the NBT and, for future expansion, the planning commission should continue to allocate funds as it would for any other activity to promote development. This could be regarded as a national investment.

But even to meet their running expenses, the radio and television services will require assistance from the government. It is considered that a deficit of about seven million dollars per annum would have to be made good by a government subsidy for about five years. This is not considered inappropriate because the broadcasting services would be used by the government for information and instruction about many of its plans and policies related to nutrition, health, family planning, agricultural productivity, non-formal education, and so forth. By the end of five years the NBT should be able to generate sufficient revenue through licenses and

Fig. 11-4. Gangubai Hangol, a distinguished singer accompanying herself on an Indian instrument.

commercials and from the sale of time on its transmitters to all the agencies which want to use broadcasting for their campaigns for economic progress and social change.

Among the major recommendations is the priority given to broadcasts for rural listeners, hitherto neglected by the radio and television services which are largely urban in location, staffing and programming. The working group, very sensibly, suggests that as between radio and television there can be no question that the former should the medium to communicate with rural India for many years to come. The technology is in every way better suited to the needs of communication with rural listeners in an underdeveloped society than the much more sophisticated and expensive technology of television.

The report reiterates the oft-expressed view that radio programs for rural listeners must be broadcast from low-power stations located in the countryside, staffed partly by professionals and partly by volunteers from the area. On the basis of the experience of community radio in other lands, the working group is confident that local or community radio can bring about substantial changes in the tradition-bound Indian society.

Besides these major suggestions, the working group has also tackled a number of matters of considerable import to professional broadcasters. They have proposed changes in the methods of selecting and training the staff, improving the conditions of service, greater autonomy for the staff at the stations and less direction from New Delhi, the reorganization of audience research, and the restructuring of educational and instructional broadcasts to make them effective.

The group's proposals have been widely welcomed both by professional broadcasters and by others. But there are doubts and difficulties.

217

The first of these is based on the assumption that having gotten control over a powerful medium like broadcasting, the Government of India would not grant *real* autonomy and the trustees may be so selected as to submit to pressure from the government. There is also the fear that parliament will not confine itself to broad questions of policy but will discuss the day-to-day working of the NBT.

Even when a government grants autonomy to a broadcasting service, the temptation to interfere with it is ever present. Such a well-established and respected organization as the BBC has to resist the government's attempts to curb its freedom. Other autonomous organizations in Canada, Australia, in Western Europe and elsewhere face even greater problems in this respect. The chances of such interference are much greater problems in this respect.

The chances of such interference are much greater when there is a monopoly than under competitive conditions. Many critics of the report argue that genuine autonomy can flourish only in a competitive system in which all the media are independent of government. There must be alternative sources of programs, especially news and comment, to ensure that all points of view can find expression. A huge monolithic system, eventually controlling about 600 radio stations and an increasing number of TV stations, would have far too much power. Smaller competing units would provide greater choice for listeners and viewers.

It is also argued that it is inconsistent with freedom of information to place in the hands of one body the power to decide what should be broadcast and by whom. As a monopoly, the trust would have tremendous power of employment and engagement. Staff and outside contributors must conform to the trust's ideas of what is good broadcasting or else they would cease to be seen or heard on TV and radio. A uniform approach to news is undesirable in a democracy. A variety of sources of editorial judgment is essential in a free broadcasting system.

Another point of criticism is the trust's dependence on government for some time for its funds. This would affect its independence. An alternative is to ask the government and all its departments and agencies which use broadcasting to carry their messages to pay for time on the transmitters, just as any commercial concern would be expected to do. Whether parliament, when it considers the matter and passes the act to change the present system, will take note of the hopes and fears expressed in the press and in public debates remains to be seen.

What are the chances of success in transfer of authority from the government to an autonomous trust? There are two political considerations. One is ideological; the present government is committed to the idea of freedom of speech and information. The other, no less valid, is the recognition that in a parliamentary democracy, with two major parties forming the government in succession to each other, the control of the media by government is of not much use in the long term. Two parties can play at the game of misusing the broadcasting service.

Ultimately, the success of the autonomous service will depend on whether the government, parliament, the public and the broadcasting

218

Fig. 11-5. A program for rural women being broadcast from the Jaipur station of All India Radio.

profession behave with responsibility. Over a period of time, a balance will have to be achieved between the interests of the government, the prerogative of parliament, the needs and wishes of the public and the freedom of the broadcaster to devise programs according to his professional judgment. Such a balance has not been easily achieved even in more mature democracies than India's. It demands a commitment to democratic principles and practices in which the good of society as a whole prevails over personal and sectional interests. But the realization that full autonomy can only be ensured in a favorable political, idealogical, and psychological climate does not mean that such autonomy should not be granted immediately. The process of adjustment and understanding needed for success should start as soon as possible.

After 40 years of stagnation, the Indian broadcasting service is thus set for reform and reorganization. The Indian public, used over the years to a service run by the state, with all the defects which stem from such control, can now hope for a service independent of the government and responsive to public opinion. If the hope is fulfilled, India will be unique among developing countries to have a broadcasting service appropriate to a democratic society and the voluntary handing over of authority to nonofficial hands will establish a democratic precedent of which the government of India could justly be proud.

Chapter 12
Broadcasting
In The Federal
Republic Of Germany

Dr. Richard Dill
Coordinator of International Relations
Rundfunkanstaltender Bundesrepublik Deutschland

The first public program in Germany was broadcast in Berlin on October 29, 1923. Before that there was only broadcasting for the military and for some news agencies. Because of a 1892 law, the German Postal Service had, from the beginning, a monopoly for the broadcasting of radio signals and was able to force a chargeable monthly license.

HISTORY OF THE GERMAN BROADCASTING SYSTEM

In 1923-24, nine regional radio corporations were founded which united to a "Reichsrundfunkgesellschaft" (Broadcasting Corporation of the Reich) RRG in 1925.

The Postal Service of the Reich got 51% of the original capital, 49% of the shares were private property. In 1926, a high postal official, Hans Bredow, who had worked in industry before that, was appointed broadcasting commissioner of the Reich by his secretary.

In 1932, the Von Papen government decided to transfer the shares of the private stockholders to the federal states. With that, the broadcasting system became the state broadcasting system, which Goebbels could take over and centralize without any difficulties in 1933. By 1945, broadcasting was an instrument of the mass-propaganda service of the national-socialistic state.

After 1945, the occupying powers called the broadcasting system, according to their own conceptions, in their occupation zones into being again. The centralized RRG was superseded by decentralized regional radio corporations.

POSTWAR BROADCASTING

Neither postal service, the government, nor private business interests should be able to control the broadcasting system. Because of that the "Anstalt des Offentlichen Rechts" (Institution of the Public Rights) was founded, which is still effective today. A system of more or less independent councils attends to the interests of the general public. A managing board assists the superintendent in managing his official duties.

Management of the corporation is in the hands of a superintendent, chosen periodically by the councils. Between 1948 (NWDR) and 1956 (SR) the "Besatzungssender" (occupation stations) became, by means of agreements of the German Federal States, today's federal-state radio corporations.

Consequently, the regional radio stations already existed at the foundation of the Federal Republic of Germany in 1948. The constitutional law let them in the cultural field, thus within the legislative responsibility of the federal states.

At the same time, as the postal service of the Reich had previously, the federal government took over the broadcasting monopoly which according to German law also defines broadcasting as the distribution of signals "langs and mittels eines lieters" (along and by means of a conductor).

In 1950, the federal-state radio corporations united to a union, whose combination of letters ARD (Arbeitsgemeinschaft der offentilich - rechtlichen Rundfumkanstalten der BRD, the community of the public, legal radio corporations of the Federal Republic of Germany) still today is the sign of the first TV channel. Today the nine federal-state radio corporations belong to it, as do the two radio corporations founded in 1960, which are transmitting for Europe and the world: Deutschlandfunk (DLF) and Deutsche Welle (DW). One of the superintendents is elected president for two years.

Television was reintroduced on a national basis within the scope of the ARD in 1954. Test programs had been broadcast since 1936. The ARD co-program started on Nobember 1, 1954.

The federal government, under Konrad Adenauer, tried to install a national TV program under a co-arrangement of government and the federal states. The German federal states successfully brought an action against this intention. The supreme German court, the federal constitutional court, denied the federal government the right to cooperate in the arrangement of the TV program in its "Fernsehurteil" (television judgment), proclaimed on February 28, 1961, and formulated these for the German broadcasting system, which are still effective today.

Instead the German federal states contracted a treaty in 1961 concerning the establishment of the ZDF, of a new corporation in Mainz, which has been broadcasting a second national program since April 11, 1963, and which is also financed by charges to the subscribers.

Out of the 11 members of the ARD, the first two (DLF, DW) are broadcasting only radio programs. The others are broadcasting both radio and television programs. This section deals mainly with and concentrates on the television performance of the broadcasting corporations. Answers to the following questions will reveal much about the performance of the television system in West Germany:

- What is programming?
- Who gives the license?

- Who controls it?
- Who finances it?
- Who plans, arranges and produces it?
- What elements comprise programming and how are they composed?
- Who is watching it and how is it accepted?

WHAT IS PROGRAMMING?

In consideration of German programming reality of the year 1980, program is the visible and audible performance of the public-legal broadcasting corporations in the Federal Republic of Germany, which is the scope of the legislature of the federal states. By means of public dissemination program is accessible to 22.2 million permanent subscribers for radio broadcasting (and about 20.3 million for television broadcasting, July 1, 1979) for a monthly charge. Ninety-three percent of all households have a TV set, 50% of all households are able to receive color TV programs.

Program is a performance of the society. It is produced and transmitted by order or by toleration of the society for the society. This all-social program regulation is primary. Broadcasting laws, program directions and other regulations of the program regulations derive out of this.

Program in the Federal Republic of Germany is based on the definition of our country as an industrial state and as a parliamentary democracy. Program regulation in the democracy is pluralistic, i.e., it presupposes a parallel and competing variety of cultural values and political opinions. The term "industrial state" points out that the society is able to afford this variety and plurality, i.e., that it is able to develop both the technical-economic hardware potential and the creative software potential for the realization of the programs.

Formally, the social program order is reflected most clearly and intelligibly in the broadcasting laws. The classical program regulation in the western democracies assigns to the broadcasting system the task to inform, to educate and to entertain.

This classical triad of programming responsibility reminds one of the origin of television: on the one hand it is considered as the extension of the public-medium newspapers (information), on the other hand as the technical daughter of theater, film and vaudeville theater (entertainment) and finally as the supplement of school and adult college (education).

In parliamentary democracies, which convoke and recall their governments by means of public and general elections, the public interest has especially shifted to information and opinion programs in the last years. Sir Charles Curran, for many years program director of the BBC, once accordingly assigned to this part of the program the task to convey to the spectator the controversial opinions of those groups, which apply for the government, and to put the spectator into the position to come to a competent judgment of the competing offers at the ballot box. Superintendent Von Sell (WDR) considers the program as "a contribution to the formation of will and decision in the society" and "the functioning of democratic processes."

Politicians more and more believe, even if it is not provable, that television programs decide or partly determine how the citizen votes, i.e., he votes on whether governments keep or lose their majorities. This has led in all states, but above all in cases where slight vote fluctuations can cause a change of power, to situations in which there are direct attempts to influence information programs by powers whose destiny is dependent upon the outcome of elections.

THE LICENSING PROCEDURE

The Federal Republic of Germany has bound itself to a comprehensive conception of freedom of thought and of information in Article No. 5 of its Constitutional Law.

So far program is mainly dependent on wireless transmission or on signal frequencies whose range is limited, and whose allocation and use, therefore, is subject to international arrangements. This calls for distributional competence of the government.

The sovereignty of telecommunications results from the frequency distributing competence of the federal government, formulated in the federal states in response to the broadcasting charge of 1969. This means that each kind of program distribution, if it is broadcast over the air or along and by means of a conductor, if it is by terrestrial transmitter, by satellite or cable, if it is for the general public or for a limited publicity, has to be licensed by the federal government, represented by the German Federal Postal Service.

At the same time, the material which is carried on these postal channels is ranked among culture and thus belongs, as education, to the regalities of the federal states. Therefore, all corporations of the ARD and the ZDF owe their existence either to broadcasting laws of a federal state or to agreements between several federal states (ZDF, NDR).

With the establishment of the corporations and with licensing authority the federal states have undertaken the commitment to provide funds to finance the corporations which they have set up. Also, the amount of the monthly subscription charge is laid down by the diets and thus strengthens their competence for guiding rules.

The critical question concerning future program development is if either, with the availability of more distribution channels, the obligation for license for new corporations is dropped or if it is possible to persuade or force the license distributor to license new corporations besides the existing.

Whether or not the existing conception of broadcasting is to be valid further on (i.e., if the traditional corporations will keep their program monopoly) is at present being tested with the organization of pilot cable projects. These pilot projects are, according to the recommendations on an independent commission of experts (K & K), to be carried through in the early '80s in some federal states and are to lead to a judgment formation about the installation of cable television.

Concerning the future of the electronic media in the Federal Republic of Germany, two clear fronts have formed: the social-liberal parties SPD and FDP intercede in common with the trade unions and the churches for the maintenance of the public-legal system. The conservative parties CDU/CSU on the other hand, do not want only to enforce private economic institutions in the program field but also in the distribution area.

Thus, the future of broadcasting, the licensing procedure and the question of who is going to be the holder of the institutions, are determined by the election results of this year.

BROADCASTING CONTROL

Control means that there are instances where checks are made to determine whether or not the program organizer fulfills his commission. According to the broadcasting laws and to the television judgment of the federal constitutional court in 1961, control is with the social-relevant powers. The representatives of these powers are organized as broadcasting councelors on broadcasting boards, where they represent the "interests of the general public." In 1961, the highest German court opposed attempts to avoid this system.

It is up to the legislator to decide who belongs to the social-relevant powers. This function to select gives a strong political influence to the governmental parties in the diets.

In two broadcasting corporations, in the NDR and in the WDR, there is an advisory board besides the broadcasting councilors. Besides that, an advisory board to which one representative out of each corporation is assigned advises the program director of the German television on questions about the ARD co-program. This board has nine members.

The councilors are not authorized to enforce or to drop programs. Their most important instruments are budget power and the power to elect the superintendent and to co-determine the appointment of executive employees.

At the same time, the right of appeal and the right of agreement of the bodies concerning the staff have increased. It no longer refers only to the appointment of the superintendent and the classical directors (program, administration, technology, law). In the BR the right of agreement of the broadcasting board refers to the appointment of executive employees, who, in addition to that, have to be confirmed in their posts every five years. In the course of the performance of this regulation it can be observe that the circle of executive employees is extended continuously, which means that the political influence on the staff of the bodies more and more interferes with the middle of the staff structure.

Bodies are more and more seeking direct influence on program decisions. In April, 1977, the administration board of the NDR declared with the majority that a sector of coverage (about the building and opening of the Brokdorf nuclear power plant) was in contradiction to the constitution, a decision against which the superintendent has appealed in the orderly courts.

The fact that the superintendents in five out of 10 German corporations have already been elected out of the control bodies and that there is a skilled writer in only one corporation, who is at the head, underlines the politicizing of the election of the superintendent and the shifting of the control from the external supervisory bodies to the corporations themselves.

This politicizing finds counterbalance in the demand for codetermination by the employees, even in the bodies (exemplary for that is the new broadcasting law of Radio Bremen) which give the employees an interest in the advisory boards of the corporation, especially after editorial co-determination.

The control from below by the employees, by the representation of the employees, is gaining increasing influence. Industrial law regulations ensure the freelance contributors, after certain regular occupation, lifelong permanent employment, even against their will, and resulting out of that the restrictions with the employment of freelance employees, strikes and efforts of the trade union to limit foreign program imports (in England program imports are limited to 14% of the program import) or, as well as in England, to limit replay and restrict the free play of the program.

FINANCING

Television in Germany is financed both by receiving set charges and by advertising receipts. In the corporations of the ARD, advertising amounts to about one third; in the ZDF (after the increase in charges on January 1, 1979) to 38%, after having been almost 50% before that. Besides that, in the federal republic there are precedents for governmental subsidizing of air programs, like the promotion of the Federal Department for Education and Science for preschool children.

The combined monthly charge for radio and television is DM 13.00 ($7.20); DM 3.80 ($2.10) for radio and DM 9.20 ($5.10) for television. Thirty percent of the television charges belong to the ZDF, which has to meet occuring financial deficiencies by advertising; advertising is limited to 20 minutes on six days a week.

The revenue received by the German broadcasting corporations amounts to three billion marks (1.7 billion US dollars) a year. Together with advertising receipts of the ARD (922 million marks, $512 million, in radio and television in 1977) and of the ZDF (340 million marks, $184 million, in 1977, approximately 394 million marks, $219 million in 1979), broadcasting in Germany turns out to be a turnover giant in the 4 billion ($2.2) range, which holds one of the first places in a list of the biggest media groups in the world beside CBS, RCA and ABC.

In 1977, the ZDF spent 890 million marks ($495 million) for the organization and transmission of television programs all over Germany. In 1979 its receipts from charges (considering the increase realized) were expected to be 637 million marks ($354 million) and the receipts from the advertising will be 393 million marks ($211 million). For the first time the range of one billion($0.55 billion) will be passed.

In 1976, one program minute cost an average of DM 3206 ($1781) in the ZDF, of which the mere production costs (which the ARD doesn't establish) totaled DM 1649 ($916). The comparable global and minute costs of the ARD are more difficult to find out, because the ARD program is divided. The ARD program cost (without playback and transmission costs) was 526 million marks ($293 million) in 1977, which corresponds to a minute price of DM 1492 ($829). This average price absorbs both the minute price for expensive lines (television play: DM 6480, $3600 per minute) and less costly lines (sports transmissions: DM 1845, $1025 per minute).

To find out the total costs for the first program (ARD), the regional ARD programs have to be considered. In 1977, they cost 313 million marks ($174 million), which corresponds to a minute price of DM 1492 ($829). These programs are financed out of the yield of the regional advertising branches, which reached gross returns of 671 million DM ($373 million) and which delivered net profits of 81 million DM ($45 million) to the corporations. If you add the program before noon, the playback and transmission, and also consider the profit of the television advertising branches, the ARD program required 1.14 billion DM ($0.63 billion).

Though the costs of the third programs permit few conclusions to the cost of new programs because of the special structure and composition of these programs, the cost of the programs recorded here are for those in 1977 (without playback and transmission): they are 391 million DM ($217 million).

Everyone who wants to advertise in the total ARD range has to pay about DM 120,000 ($67,000), in the ZDF, DM 83,400 ($46,300) in 1979.

In 1977, the ARD had 18,085 permanently occupied employees (without DW and DLF), the ZDF had 3378. In 1977, the staff expenditures of the ARD (including pensions and provisions for pensions) totaled 1.3 billion DM ($0.72 billion); those of the ZDF totaled 220 million DM ($122 million). With that the quota for staff costs has reached 25% in the ZDF, while in the ARD corporations it is drawing near the 50% range or is passing it; staff costs in the NDR in 1976 amounted to 56%).

THE PROGRAM ORGANIZERS

By the beginning of 1980, the German television spectator will be able to choose between four domestic television programs, which will be offered to him by two organizers on three channels, by the nine ARD corporations and by the ZDF:

● the national co-program of the ARD
● the ARD regional   program on the first channel on weekdays between 6 p.m. and 8 p.m.
● ZDF program all over Germany
● one of the five third channel programs, which are offered by the federal corporations alone or cooperatively with neighboring corporations.

Besides that, additional domestic or foreign programs are accessible to almost all spectators.

226

The ARD, a union of broadcasting corporations with only limited central competences and organs, was founded in 1950. In 1953, the ARD members made a television agreement:

"The German television program consists of the program contributions of the contracting broadcasting corporations."

According to their charges levied the nine members contribute the following percentages to the ARD co-program:

WDR	25
NDR	19
BR	17
SWF	9
SDR	8
HR	8
SFB	8
RB	3
SR	3

This program quota has a character of obligation and of right. For example, a central ARD program management (with residence in Munich) is able to demand the portion of the program and that for all lines of the program separately and for all program places. The corporations can claim the program places which they are entitled to. This form of organization for the program led already before the installation of the ZDF to the "Kästchenprogramm" (little box program): not the contents but the competence is the primary decisive motive. It is not important what is being transmitted but who is entitled to which program place. Therefore, the discussions about the programs within the ARD always mean only the discussions of structure places, rarely of forms and contents, in spite of certain attempts to counteract.

With the appearance of ZDF appeared in 1963 there came into the world singular—and probably unrepeatable—schematizing of the program offers. For the federal states gave the order to the new central corporation to transmit a program differing in contents. Subsequently, the existing federal broadcasting corporations were also liable to this so-called contrast program. Formally, both systems fulfill the instruction by taking a general agreement about the program for a limited time and by agreeing on a program scheme, which is based on a common though controversial definition of contrast. The present program scheme has been in operation since January 1, 1978.

But the obligation to coordinate only refers to the main transmission time from 8 p.m. on and can, therefore, be restricted to three contrasting formats:

- information
- play and movie
- entertainment

Regional programs, third programs and also the programs in the afternoon and on the weekends are excepted from this coordination.

WHAT DO PROGRAMS CONTAIN?

In 1977, the ZDF offered program for about 200,000 minutes, which corresponds to a daily average of nine hours and 14 minutes. The ARD co-program transmitted during the same period (it begins its version of access time from 6 p.m. to 8 p.m.) was about 172,000 minutes. In addition to that, there are the regional programs with about 35,000 minutes each. In 1977, the annual volume of the third programs was between 125,000 minutes (Nord III) and 188,000 minutes (WDR).

The most important criteria for the quantitative description of the programs are: 1. contents and 2. sources of the programs. The sorting of programs according to contents depends on the program regulation. With that the basic division into three parts is on the one hand simplified and on the other hand differentiated.

Under the viewpoint of the administration of the program the ZDF, for example, has installed two program responsibilities: a chief editor's office for all current political and documentary broadcasts and a program management for all the other fields.

The ARD's conception of its division of program contents can be seen by the divisions of competence of the so-called program coordinators, which combine the individual performances of the ARD corporations to the national program. Under one program director the ARD has nine coordinators. The coordinator for the family-time program cares for the structure places in the afternoon before 6 p.m. for children, juveniles, women and seniors and with that 13% of the co-program (irrespective of contents). The coordinator for foreign programs is dealing with international interchange of programs in both directions and yields from the Eurovision between 2% and 4% of the co-program.

The coordinator for film promotion cares for the accomplishment of the agreement between ARD (and ZDF) and the corporation for film promotion in Berlin, in the scope of which about 44 million DM ($24.4 million) of ARD money was provided between 1974 and 1978 for co-productions with film companies. The other six coordinators are dealing with the filling of structure places in their respective lines, as listed in Table 12-1.

The questions for the sources of programs are: 1) how much program is produced by the corporations themselves (either in the studio or as a film) and 2) which quota is bought or taken over? For 1977, the ZDF reports a quota of its own production of 54.8%.

The criteria for the ARD co-program are different because of the different formation of the program. Here the main criterion is the question: is a certain broadcast credited for the fixed quota of the program or not.

The origin of the program plays a secondary part. It is left to the corporation if it fills its program places with its own productions or with foreign productions (foreign in the sense that the productions come from other corporations in Germany or from other countries).

The second ARD criterion distinguishes between performances of the single corporations and so-called co-programs. To the co-programs belong on the one hand all programs, which are commonly bought and sold

228

(for example, movies, Eurovision broadcasts, transmission of Olympics). On the other hand, so-called national delivery programs, to which all corporations contribute, for which the compulsory portion formula isn't valid. News and sports news, for example, belong to that. In 1977, the relation between contribution of the corporations and co-contributions was 70:30.

In opposition to many foreign program statistics, ARD and ZDF don't give any information as to how they use foreign program sources. At present, there are no regulations for the restriction of foreign parts of the program as they exist in other countries.

As a third criterion there is to record the quota for replays. It is important because the right for replays is restricted in more and more countries. At the same time, the idea of structurized replays plays an important part in the planning of future programs. Pattern for a future replay program is the ARD/ZDF program in the morning, which in the first place is transmitted by the corporations neighboring the GDR for GDR spectators but whose expansion to the whole Federal Republic is considered. In 1977, the quota of replays in the main programs totaled 13.9% for the ZDF (chief editorship field 1.1%, program management field 22%) and for the ARD 14.4%.

AUDIENCE RESEARCH

Each program organizer must ask himself the question, how is he faring with the audience for whose joy and use the whole television business is organized? The commercial answer: "We give the spectators what they want," and the reversal belonging to it, that each other program is dictatorial, arrogant, holding in leading strings, undemocratic, more and more corporations answer with the statement that in the program the spectator also should be offered what he needs.

Table 12-1. Program Quotas Offered German Audiences in 1977.

PROGRAM CATEGORIES		PROGRAM QUOTA IN 1977 (%)
Politics, society and culture	Documentary Broadcasts	22
	News and Weather	10.1 = 32.1
Sports		7.3
Play		10.4
Entertainment		16.2
Movie		8.6
Church broadcasts		1.4
Music		0.8

Table 12-2. Network and Station Call Identities.

ZDF	-	Zweites Deutsches Fernesehen
NDR	-	Norddeutscher Rundfunk
BR	-	Bayrischer Rundfunk
SWF	-	Südwestfunk
SDR	-	Süddeutscher Rundfunk
HR	-	Hessischer Rundfunk
SR	-	Saarlandischer Runkfunk
RB	-	Radio Bremen
SFB	-	Sender Freies Berlin
ARD	-	Arbeitsgemeinschaft der Offentlich-rechtlichen Rundfunkanstalten Deutschlands
WDR	-	Westdeutscher Rundfunk

The public legal system seems to be most suitable to realize a program compromise with regard of the spectator's taste: a mixture between desired and the indispensible.

Basic requirement for "publikumsnahe" (close to the spectators, i.e., popular) program planning is exact and differentiated knowledge of the viewer and of his behavior. ARD and ZDF both have a comprehensive research system, partly in its own management by media reports (ARD) respectively by a department of media research (ZDF), partly by common placing of orders. Since 1975, the TELESKOPIE company, a branch for the institute for opinion poll in Allensbach and of the institute for market research INFAS in Godesberg, have been investigating the daily reaction to ARD and ZDF programs by electronic questioning of about 1200 households. (Network and station call identities appear in Table 12-2.)

In the scope of the research of the viewers' quota they especially ask:

● Who the spectators are (age, sex, profession, income, education, living area and so on)?

● How the spectator is planning and spending his time of day and time of life.

● What, how and with what degree of agreement he is watching?

● How much time and attention he is devoting to the particular media offerings?

● Which long-term and short-term effects the viewed programs have on him.

The superabundance of investigations and accessible data reveal that the average viewer in Germany over 14 was watching 147 minutes on the average during the first quarter in 1978. During this time he was watching 69 minutes of the ARD program, 65 minutes of the ZDF program, 11 minutes of the Third Programs and two minutes of the foreign programs.

This average daily time in minutes the average viewer spent watching television is:

1964:	70
1970:	113
1974:	125
1976 (1st quarter):	148

Thus, the consumption of television has remained almost unchanged since 1976, in spite of considerable program expansions. Compared to other countries, the German consumption of television is low.

The investigation of watching behavior is especially important where there are not only national programs. A special TELESKOPIE investigation proved that German viewers who can receive more than three programs don't extend their daily television consumption. If you assume that the viewer likes best of all what he is watching most of the time, definite preferences for program acceptance are movies and play actions of any kind, including thrillers and western movies, sports entertainment programs, quiz programs and news.

Employees and persons responsible for the program justly think that the independence of the viewers' quota is one of the most important and justifying achievements of the public-legal television.

Chapter 13
Broadcasting in Britain

Prepared by the
Central Office of Information
British Information Services

Broadcasting in Britain, both of sound and vision, is regulated by the Wireless Telegraphy Acts of 1949 and 1967. These acts prohibit the sending or receiving of radio communications unless under license. Since 1971, the receiving of radio signals from authorized sound broadcasting stations or from licensed amateur stations has been exempted from the need for a license.

ORIGIN AND DEVELOPMENT

Early recognition of the importance of broadcasting imposed responsibilities on the BBC (and later the IBA) in such matters as the treatment of various subjects, particularly current affairs, and the choice of programs broadcast. The intention has always been to maintain as far as possible the independence of the broadcasting authorities while ensuring that they retain a balanced approach to their responsibilities. Periodic committees of inquiry, which have reported either on a particular aspect of broadcasting or on the subject as a whole, have reflected the need for constant review of the role of the broadcasting authorities and for change in response to new developments.

British Broadcasting Company

When broadcasting began in Britain, parliament decided that the postmaster general should grant only one license for that purpose at any one time. (The office of postmaster-general was abolished in 1969 and that of minister of posts and telecommunications created. In 1974, the ministry of posts and telecommunications was abolished. Its responsibilities in relation to the post office were transferred to the department of industry and its function in the field of broadcasting to the home office.) In 1923, the first license was granted for two years exclusively to a limited company, the British Broadcasting Company, which was formed by combining the broadcasting interests of six large wireless and electrical manufacturing firms. The company at first derived its revenue partly from royalties charged on the sale of wireless receiving sets and partly from the sale of receiving licenses by the post office. Transmitting stations were opened in London, Manchester, Birmingham, Newcastle upon Tyne, Cardiff and Glasgow.

The company's license was continued for a further period of two years after an official committee of inquiry (the Sykes Committee) had recommended that, though the ultimate control of broadcasting must rest with a minister responsible to parliament, the state itself should not operate the broadcasting stations but should establish a broadcasting board to do so; and that consideration should be given to extending the broadcasting system by the establishment of a number of local or relay stations, so that wider audiences might be reached. This the company undertook to do, and by the end of 1926, when its renewed license expired, a national system of broadcasting stations had come into existence, and the number of receiving licenses had risen to over two million.

The Foundation of the BBC

From the beginning, it was apparent that the British Broadcasting Company, under the chairmanship of Lord Reith, appreciated the significance of its work. A further committee, appointed in 1925 by the government to make recommendations for the constitution of a national broadcasting service, rejected both the idea of free and uncontrolled broadcasting by companies operating for profit, and the idea of service directly controlled and operated by the state, as being, in different ways, unsuitable. It proposed that broadcasting should be conducted by a public corporation acting as a trustee for the national interest and consisting of a board of governors which should be responsible for seeing that broadcasting was carried out as a public service. Lord Reith supported its views. The British Broadcasting Corporation (BBC) was established by Royal Charter in 1927 and the whole staff and equipment of the former company were taken over intact. Lord Reith was appointed director general, a position which he held for 11 years.

The BBC's Relations with the Government

The BBC began its work under the authority of two documents, each granted for a period of ten years: (a) the charter, which closely followed the recommendations of the Crawford Committee and prescribed the objects of the corporation, the manner in which its functions should be exercised, its internal organization, and the financial arrangements which should be made, and (b) the license and agreement between the corporation and the postmaster general.

These documents permitted the corporation to establish broadcasting stations, subject to the postmaster general's technical conditions on wavelength and power used; they also prohibited the BBC from broadcasting commercial advertisements or sponsored programs, and defined the percentage of the license revenue which the corporation would receive as income to enable it to do its work.

Theoretically, the 1927 charter and license and agreement gave the government full powers over the BBC. Under the charter it was possible for the governors, appointed by the Crown to constitute the corporation, to be removed at any time by order in council and to be replaced by others. The license could be revoked by the postmaster-general if at any time the corporation, in his opinion, failed in its duties. The postmaster-general could veto any proposed broadcast. Any government department could require the corporation to broadcast any matter as it desired.

In these conditions, it might have seemed that the BBC could develop only as a subsidiary government department under another name. That this did not happen was partly due to the fact that it was the agreed policy of successive governments, accepted by parliament, to treat their powers as major reserve powers only and to grant the corporation absolute independence in the day-to-day conduct of its business, vesting it with full responsibility over general administration and program content for audiences both at home and abroad, and partly due to the personalities of successive directors-general, who have maintained political independence and freedom from commercial pressures for the corporation, and have upheld its standards of integrity and efficiency and its place in the international field of broadcasting. In 50 years of broadcasting no formal veto has ever been placed on the broadcasting of a particular item, and the only general directions now in force are those which prohibit the BBC from broadcasting an opinion of its own on current affairs or matters of public policy and from using techniques of subliminal broadcasting. The BBC is also precluded from being paid to broadcast a program or from broadcasting a sponsored program.

Radio Broadcasting Since 1927

Broadcasting in Britain has continued to be reviewed at regular intervals and on each occasion the BBC's license has been renewed for a further period. Over the years, technical developments have improved reception and brought radio broadcasting within the reach of larger audiences.

By 1939, the BBC was operating a number of shortwave broadcasts overseas, to the other Commonwealth countries in English and to some other countries in their own languages. The government made itself responsible for financing these external broadcasts and instructed the BBC on the services to be provided specifying the countries to be covered and the number of hours to be devoted to each; but the BBC retained substantially the same independence in framing program content as it already had in the domestic field. The Empire Service, as it was then called, began in 1932; the Arabic Service and the Latin-American Services in Spanish and Portuguese began in 1938; broadcasting in French, German and Italian followed in September of the same year. (See Fig. 13-1).

During the Second World War, radio broadcasting was immensely significant. The domestic services, while retaining their independence, were not only an important channel of news, national leadership and guidance, but also a pool for ideas and information, and the most readily available medium of entertainment and recreation for the great majority of people. The Empire and Overseas Services expanded into a world service of broadcasting, unequalled in its technical resources and coverage and in the range of its language services.

Developments since 1945 in the radio services include the introduction of the Third Program, designed to provide programs of a high cultural quality, and the increasing provision from 1955 of very high frequency (VHF) transmissions permitting much clearer and more accurate reception than on the long and medium waves, which are increasingly crowded by the number of broadcasting authorities making use of them. The Music Program was introduced during 1964 as a service of serious music broadcast on the Third Program wavelength during the daytime, and regular sterophonic broadcasting of music began in 1966.

A reorganization of domestic radio programs took place in the period 1967-71 with the establishment of four networks, each with a distinctive output and style. The old Home Service, Light Program, and Third/Music Program were replaced by Radios 2, 3, and 4 and a new network, Radio 1, created to broadcast continuous "pop" music. Eight local radio stations, run by the BBC on an experimental basis, began broadcasting in 1967 and 1968 on VHF and this service was established on a permanent footing when a further 12 stations were added to the network in 1970-71. The potential audience for BBC local radio increased after September, 1972, when many of the stations started transmissions on medium wave.

Fig. 13-1. The cast of a radio play broadcast in the BBC world service for Africa.

The Sound Broadcasting Act 1972 (later incorporated with the Television Act 1964 in the Independent Broadcasting Authority Act 1973) authorized the creation of independent local radio stations under the auspices of the Independent Broadcasting Authority. The number of stations is being limited to 19 pending the report of the Annan Committee of inquiry into the future of broadcasting.

Outside Britain the postmaster general, at the request of the government of the Isle of Man (and in recognition of the island's separate identity), in 1964, licensed the Isle of Man Broadcasting Company to operate a broadcasting service confined to the island. The company, which was taken over in 1971 by a public body, the Isle of Man Broadcasting Commission, operates a commercial service.

As part of international action to end the illegal operation of offshore (pirate) commercial radio stations, the government promoted the Marine etc. Broadcasting (Offenses) Act in 1967, under which the stations around Britain were closed down. The government and the BBC recognized, however, that there was a need for a program of continuous popular music and this was subsequently introduced by the BBC on Radio 1.

Although there has been a reduction in the strength of the radio audience as a result of television, this has been slowed by developments such as the production of cheap, portable transistor radios which greatly increase the convenience of listening. Radio continues, therefore, to play a vital part in the broadcast services; it provides news services, music and a wide range of other programs throughout the day and has more program time than television in which to cater to minority interests of various kinds.

Television Broadcasting

Research into the technique of television was pursued by scientists of several nations from the 19th Century onwards, but its emergence from the laboratory may be said to date from 1926, when the first public demonstration in the world of true television was given at the Royal Institution of London by John Logie Baird, showing images of moving human faces with gradations of light and shade. Three years later the BBC began experiments in broadcasting television by granting facilities to a company formed by Baird (Baird Television Limited) to transmit programs through a BBC transmitter outside ordinary program hours. By 1932, these transmissions had proved of sufficient technical interest to justify the BBC in equipping a studio in Broadcasting House with the Baird mechanical system, and in creating a small organization within the BBC to run a few experimental, but regular, transmissions each week.

By the beginning of 1934, several systems, other than the mechanical one invented by Baird, had been worked out, including an electronic system (evolved in the United States and developed in Britain), which became known as the Marconi-EMI system. At this stage the government appointed a committee to report on the relative merits of the various television systems and on the conditions in which a public service might be introduced. As a result of its recommendations the BBC was made

236

responsible for television as for radio; a Television Advisory Committee (TAC) was appointed by the postmaster-general, and a television station was formally opened at Alexandra Palace in London.

In 1936, the BBC launched the first regular public high-definition television service in the world which began with the daily transmission of a 2-hour program, using the Baird and the Marconi-EMI systems during alternate weeks. Three months later the Baird system was discarded on the recommendation of the TAC, which suggested the adoption of a single set of standards for transmissions, and thenceforth the Marconi-EMI system was used to carry on the service alone.

By 1939, it was estimated that the total number of television receivers in Britain was just over 20,000. Then the war broke out and the television service was completely suspended, to be resumed by the BBC in 1946. By 1952, there were 1.5 million combined television and radio licenses; this rose to some 5.7 million in 1956 and to 10.5 million in 1960. In February 1975, there were 17.5 million television licenses. (Separate licenses for radio only were abolished in 1971.)

The Establishment of Independent Television

Within a few years, the growth of television and its increasing popularity among all sections of the community were arousing public interest, since it was obvious that, as a medium of entertainment, it was having a profound effect on the social habits of the British people. The Beveridge Committee, which reported of the future of broadcasting in 1951, recommended its continuance by a public service corporation alone subject to safeguards against abuse and monopoly. However, there were other views and one of the most important proposals in the conservative government's statement on the Beveridge report referred to the possibility of introducing an alternative television service provided by private enterprise.

Toward the end of 1953, the government published proposals designed to incorporate important elements of both viewpoints: a commercial television service would be established as an alternative service putting an end to the BBC's monopoly, but it would operate according to the principles of public service, under effective government control. These were embodied in the Television Act 1954, which established an Independent Television Authority "to provide . . . for a period of ten years from the passing of this act, television broadcasting services additional to those of the British Broadcasting Corporation and of high quality . . ." The act laid down that the actual programs should, in general, be provided, not by the authority by program contractors "who under contracts with the authority, have, in consideration of payments to the authority . . . the right and duty to provide programs or parts of programs to be broadcast by the authority, which may include advertisements".

The Independent Television Authority (ITA) was set up in August 1954; the first regular television programs were transmitted in September, 1955. (See Fig. 13-2).

Developments Since 1960

The experience of the first few years in which the BBC and the ITA both provided television services was reviewed, together with the BBC's radio services and possible future developments, by a further committee on broadcasting, the Pilkington Committee which reported in 1962. The principal recommendations accepted by the government were: the BBC should remain the sole provider of radio services, the hours of which would be extended; the BBC should provide a second television service which should be in the ultra-high frequency (UHF) bands, on a standard of 625 lines (the internationally agreed standard in use throughout most of Europe); existing 405-line television services should eventually be duplicated in 625 lines; and color should be used (on 625 lines) as soon as possible. All of these proposals have been implemented: the BBC's second television service, BBC 2, started in April, 1964, and in 1967 the first regular service of color television in Europe began on BBC 2.

Color was introduced simultaneously on BBC 1 and Independent Television in 1969. All color programs are transmitted on 625 lines on UHF and the transmitting stations are designed so that (with minor exceptions) each station can broadcast all three of the existing television services and could readily be adapted to broadcast a fourth service if required. By February, 1975, there were 7.3 million households with licenses for color television.

The committee made some proposals for Independent Television, but these were in the main not accepted; the government decided to retain the existing structure, and to give the ITA certain additional powers, particularly over programs and advertising.

The Crawford Report

In 1973, the minister of posts and telecommunications announced the setting up of a Committee on Broadcasting Coverage under the chairmanship of Sir Stewart Crawford. At the same time the minister separately invited bodies, including the broadcasting authorities, to express views on the introduction of a fourth television network to supplement existing services.

The report of the Crawford Committee was published in 1974. The main recommendation of the committee, accepted in principle by the government, was that the fourth television channel in Wales should be allotted to a separate service in which Welsh-language programs should be given priority and that this service would be introduced without waiting for a decision on the use of the fourth channel as a whole. The report recommended that the service should be provided on a joint basis by the BBC and Harlech Television and that a government subsidy should be considered. The home secretary set up a working party to consider the implications of implementing this recommendation. Another recommendation accepted by the government was that the extension of UHF coverage throughout Britain should take first priority.

Committee of Inquiry into the Future of Broadcasting

In 1974, the government set up a Committee of Inquiry into the future of broadcasting under the chairmanship of Lord Annan, provost of

Fig. 13-2. The set of a drama series being produced by Granada TV, one of the independent TV companies.

University College, London. The committee's terms of reference were "to consider the future of the broadcasting services in the United Kingdom, including the dissemination by wire of broadcast and other programs and of television for public showing; to consider the implications for present or any recommended additional services of new techniques; and to propose what constitutional, organizational and financial arrangements and what conditions should apply to the conduct of these services." To allow time for the committee to complete its task the government extended the charter of the BBC and the Independent Broadcasting Authority (No. 2) Act similarly extended the life of the IBA.

Broadcast Relay System

Systems whereby broadcast programs are received at a central point and distributed to homes by wire or cable began in Britain in 1925 as a private venture, and they remain largely in the hands of private enterprise.

Broadcast relay systems operate under license from the home secretary and the post office. Their function is to relay BBC and IBA programs from broadcasting stations and in general they do not originate programs of their own. In 1972, however, an experiment was authorized by the government whereby a number of relay companies have been allowed to originate and transmit locally produced television programs.

Subscribers to cable television broadcasting services also need ordinary television receiving licenses.

In 1973, there were about 1950 systems working under 1285 licenses. Some 13% of all television households obtained their reception of broadcast programs from a broadcast relay system. The number of subscribers was about 2.3 million, some of whom received only radio programs.

ORGANIZATION OF BROADCASTING AUTHORITIES

The BBC and the IBA are the two authorities for regulating all broadcasting in the UK.

British Broadcasting Corporation

The BBC consists of 12 governors (including a chairman, a vice-chairman and national governors for Scotland, Wales and Northern Ireland), each appointed for a period of not more than five years by the Queen on the advice of the government. As a corporate body the governors are responsible for the conduct of the whole broadcasting operation, including the content and presentation of programs in radio and television, and the provision and working of the necessary installations and equipment.

The National Broadcasting Councils for Scotland and Wales were established in 1952 to control the policy and content of the radio programs provided primarily for reception in their respective countries and were later given comparable duties and responsibilities in respect of television. The chairmen of the councils are the corporation's national governors representing the area concerned. The national governor for Northern Ireland chairs the Northern Ireland Advisory Council.

The corporation is advised by (a) the General Advisory Council of about 60 members, which was first appointed by the BBC in 1934; (b) the eight regional advisory councils for the English regions and the regional advisory council for Northern Ireland; and (c) a number of councils and committees not specifically required by the charter but established by the BBC to advise on religious broadcasting, charitable appeals, the social effects of television, music, agriculture, schools broadcasting, further education, programs for immigrants, science and engineering. The members of all these councils and committees are appointed by the BBC for a period of four or five years at different times so as to provide for a change of membership with general continuity.

In 1971, the BBC set up a 3-member Program Complaints Commission whose task is to consider complaints of unfair treatment when the BBC answer has failed to satisfy a complainant who believes himself to have suffered unfairly as a result of a program.

The chief executive officer of the BBC is the director-general, who is appointed by the governors and with whom they discuss all major matters of policy and finance. Under the director-general are the managing directors for television, radio and external broadcasting and the directors of public affairs, engineering, personnel and finance, who, with the director-general, constitute its board of management. Under the directors come the divisional controllers and the heads of departments. The total full-time staff employed by the corporation in March 1974 was nearly 24,000.

The radio and television services of the BBC are financed from (a) an annual sum voted by parliament for the domestic services, equal to the revenue derived from the sale of television broadcasting receiving licenses after deducting the cost of administering the licensing system; (b) an annual grant voted by parliament for the external services; (c) profits from BBC publications; and (d) income from BBC enterprises.

The cost of television broadcast receiving licenses is £8 for monochrome and £18 for color television; separate licenses for radio were abolished in 1971. One license covers all the receiving sets in a household. Registered blind people are entitled to a licensed at a rate reduced by £1.25. Estimated net revenue from the sale of licenses for the year ended March 1974 was £136 million. The grant for the external services amounted to 16.7 million and the net profit made by the BBC on its trading activities in 1973-74 totalled nearly £700,000. BBC local radio costs are met out of license revenue.

Independent Broadcasting Authority

The Independent Broadcasting Authority (IBA) consists of a chairman, a deputy chairman and eight members (three of whom have special responsibility for Scotland, Wales and Northern Ireland severally), appointed by the home secretary. The IBA is charged with the duty of providing broadcasting services additional to those of the BBC in accordance with the Independent Broadcasting Authority Act 1973; it builds, owns and operates transmitting stations, selects and appoints program companies and exercises control over programs and advertising. The Sound Broadcasting Act 1972 (later consolidated with the Television Act of 1964 in the Independent Broadcasting Authority Act 1973) gave the IBA the responsibility of providing independent local radio.

The IBA is advised by a general advisory council and by three statutory committees on educational broadcasting, religious broadcasting and advertising. A specialist panel advises on medical and allied advertisements. Further advice is given by the Scottish, Welsh, Northern Ireland and appeals committees. There are also local advisory committees for independent local radio.

In 1971, the authority set up a complaints review board to strengthen its existing internal procedures for considering and investigating complaints. The board, which has four members, is concerned with complaints from the public or from persons appearing in programs about the content of programs transmitted or the preparation of programs for transmission. The chief executive officer of the IBA is the director-general; the staff is divided into eight main divisions, namely program services, administrative services, finance (external and internal), engineering, radio, advertising control, and information. Total full-time staff of the authority is 1330 and the staff of independent television as a whole (including the IBA) amounts to some 11,000 people.

The studios of ITN (Independent Television News Ltd) are designed for nationally networked news programs, and provide London facilities for the regional program companies and for overseas broadcasters. The companies provide local news programs. The whole of independent television is financed solely by the sale of advertising time by the program companies. The system receives no income from licenses or public funds.

The IBA is also responsible for independent local radio stations which are being established. The first stations, in London, began broadcasting in 1973 and it is expected that there will be 19 by the end of 1975. There are safeguards under the Independent Broadcasting Authority Act 1973 whereby a company's contract may be terminated or suspended should newspaper holdings in a program company lead to results contrary to the public interest. In practice the IBA welcomes the valuable qualities which newspapers can bring to the development of television and radio, but is opposed to the control or ownership of any program company by any single newspaper or press interest.

The finance of the IBA is drawn from annual rental payments made to it by the program companies. The Independent Broadcasting Authority Act 1973 requires the IBA to keep the finances of its radio and television activities quite separate. The rental payments are used by the IBA to provide the network of transmitters and to discharge its other functions. The Independent Broadcasting Authority Act 1973 also provides for the IBA to draw up to £2 million on loan for financing independent local radio.

The television companies also pay a levy which is passed on by the IBA to the exchequer in payment for the valuable public concession enjoyed. The amount of levy used to be related to advertising revenue, but under the Independent Broadcasting Authority Act 1974 the basis of the levy was changed from advertising to profits. Television companies pay no levy on that part of their profit which does not exceed £250,000 or 2% of their advertising revenue, whichever is greatest. For profits above these amounts the companies pay a levy of 66.7%. Since the levy was introduced in 1964, the companies have paid to the exchequer about £217 million in addition to normal taxation. A different system operates with the independent local radio companies who are not required to pay a levy as such.

Coordination between the IBA and the program companies is effected by a number of committees. The standing consultative committee and the program policy committee consist of senior representatives of the 15 program companies and the IBA. The former considers all matters of common interest to the IBA and the companies, and the latter is the main channel through which the IBA informs the companies of its views on program policy.

The work of the program policy committee is closely linked with that of the network program committee, which is the main instrument of the companies for working out the basic network schedules of independent television and which arranges cooperation between the companies on program matters; representatives of the IBA also sit on the network program committee. An important though less formal instrument of program cooperation is the program controllers' group of the five largest companies with which the IBA's program staff keeps in close contact. There is also regular consultation between the companies individually and the IBA.

THE BROADCASTING SERVICES

Until the passing of the Sound Broadcasting Act 1972, the BBC had a monopoly of radio broadcasting. Just over half the population listen to the radio on an average day but the growth of television has led to a considerable reduction in evening listening. Over the years the organization of programs has been adapted to meet changes in interest and listening habits, the most recent organization taking place in 1970.

BBC National Radio

BBC national radio services consist of Radios 1, 2, 3, and 4. Some 80% of radio listening is accounted for by Radios 1 and 2. Radio 1 broadcasts a program of "pop" music for up to 14 hours a day, including programs concerned with experimental music. Radio 2 provides a varied output of light music including jazz, brass band music, country and folk music, and is, in addition, the main channel for sport in the form of news and live commentaries. Important features of Radio 2 evening programs are light entertainment shows and drama serials.

Radio 3 offers a complete service of mainly serious music; some 100 hours are broadcast each week and it includes classical work, opera, contemporary music, experimental works, and jazz. Some 150 operas are broadcast each year including some rarely performed works. In the early evening, the network broadcasts adult education programs and the remaining time is taken up with orchestral concerts, chamber music, new plays, and drama, poetry, arts, criticism, talks and other cultural programs. It is estimated that some five million people are regular listeners to Radio 3. An additional service provided by Radio 3 is the live commentary of test cricket matches.

The most diverse radio network is Radio 4. There are four main elements, namely news, educational broadcasts for schools, general entertainment and information, and longer talks and discussions. Radio 4 is the main network for news and current affairs, some five hours a day being broadcast. During term time, the network carries over three hours a day of educational programs for schools. Programs of general entertainment include drama of all kinds, comedy shows and panel games, and some serious music. Serious talks and discussions cover a wide variety of subjects.

A popular recent innovation on Radio 4 has been the "phone-in" program in which members of the public are given the opportunity to participate in broadcast discussions usually by putting questions to speakers in the studio. For example, before the general elections of February and October 1974, the program, *Election Call*, enabled listeners to question spokesmen of political parties on their policies. Some areas like Wales and Scotland have their own versions of Radio 4, the material partly consisting of programs made specially for listeners in those areas and partly of programs broadcast on the Radio 4 national network.

A large number of radio programs are recorded on discs and tapes, and are on sale from the records division of BBC Enterprises.

The BBC also operates 20 local radio stations in England which broadcast some 9 to 13 hours of locally produced programs each day.

BBC Television Services

The BBC transmits two national television services: BBC 1 on 405 lines on very high frequency (VHF) and 625 lines color on ultra high frequency (UHF), and BBC 2 in color on UHF only. BBC 1 is available to over 99% of the population and BBC 2 to over 94%.

Although the home secretary has power to regulate the hours of television broadcasting, restrictions on the amount of television broadcasting were lifted in 1972 and both the BBC and IBA extended their hours of transmission. However, in order to effect certain economy measures, both the BBC and the IBA reduced the transmission time of some of their radio and television services in early 1975.

The wide range of BBC productions includes talks and documentary programs, drama and films, sport and outside broadcasts of all kinds, light entertainment, music, family programs, broadcasts for schools, educational programs for adults and religious programs. (See Fig. 13-3).

The planning of BBC 1 and BBC 2 is broadly complementary, with a number of junctions in programs so that the selective viewer may switch conveniently from one program to another. The BBC's second channel gives greater opportunities for the provision of programs for minority interests or specialized needs, often at peak viewing times; for example, programs of an educational nature, those for the deaf, and broadcasts of sports such as Rugby League, golf and one-day cricket matches. Extended coverage may also be given to certain programs such as the live transmission of a complete opera. (See Fig. 13-4.)

The Community Program Unit of the BBC enables groups or associations or sections of the community to make their own television programs, which are broadcast in an occasional weekly series on BBC 2 called *Open Door*. The groups are given technical facilities and professional advice by the BBC, but decide for themselves the style and content of their programs, subject to limitations of cost and the legal requirements of broadcasting. Television programs made for the Open University are transmitted on BBC 2 in the early evening during the week and repeated on the same channel at weekends. Some BBC programs are produced by three regional network production centers in England. In Scotland, Wales and Northern Ireland the BBC transmits a number of locally produced programs designed for viewers in these areas.

Many of the most successful programs, such as the drama series *War and Peace*, the cultural series *The Ascent of Man*, and a number of historical drama series have been exported all over the world. BBC Enterprises is concerned with the sale of television programs overseas and these are distributed annually to some 90 countries, including the Soviet Union and Eastern Europe. Sequences and footage from the BBC

Television Film Library are also available for export. In 1973-74, gross income from the sale of BBC Enterprises television programs was £4.3 million.

BBC External Services

The external broadcasting services of the BBC are intended to provide a link of information, culture, and entertainment between the peoples of Britain and those in other parts of the world; to report events of worldwide importance with speed, accuracy and objectivity; and generally to reflect British opinion and the British way of life. The services provide for a growing audience: with the great increase in ownership of radio sets in the world since 1955, particularly in the developing countries where other means of mass communication are not widely available. It is estimated that in 1973 there was a world total of nearly 900 million sets.

The BBC external services broadcast in English and in 39 other languages for some 740 hours each week. The services, under the managing director of external broadcasting, function in two divisions, one responsible for all the transmitted output, the services division, and the other, the program division, concerned with program production. The program division includes the external services news department which produces about 250 news bulletins daily, and the English by Radio and Television department.

In 1972, the BBC was allocated a second medium-wave frequency, and substantial increases in European coverage by the French, German

Fig. 13-3. Puppets being used for a BBC Wales TV program for young Welsh-speaking children.

and world services took place. Virtually all the transmissions of the German service became available on medium-wave frequencies and the additional frequency only by shortwave. The changes also allowed for more effective broadcasting to central and eastern Europe. BBC world service coverage of Europe increased from 36 to more than 70 hours a week.

Independent Television Service

The program companies which compromise independent television provide the programs which are transmitted by the IBA on 405 lines on VHF and 625 lines color in UHF: films, entertainment programs, plays and serials, talks, discussions and documentaries, music, sport, children's and educational programs are the main items broadcast in an average week. Like the BBC, independent television has extended its presentation of what it calls *access* programs which enable groups within the community to broadcast their views. Many companies also have programs designed to help people with their problems, for example, by informing them of their legal rights.

The Independent Broadcasting Authority Act 1973 gives the IBA considerable powers to control the content and quality of independent television programs and their selection for showing on the national network. Each program company submits its quarterly program schedules to the IBA to ensure, for example, a proper balance between serious and light programs. The IBA also seeks to ensure that programs shown in the early evening period are not unsuitable for children. Discussions usually take place between the IBA and the individual program companies on an informal basis, but in the last resort the IBA may issue a formal instruction that a program or item should be withdrawn.

The IBA generally expects that each schedule should follow certain guide-lines. These include, for example, that one-third of all material should be serious non-fiction: that each company should provide a suitable proportion of programs calculated to appeal specially to the tastes and outlook of viewers in its area; that no more than 14% of the programs, averaged over quarterly periods, should be of imported foreign material; that on every weekday there will be an hour of programs for children shown before 6 p.m.; that between two and a half and three hours of religious programs should be shown each week; and that a minimum of nine hours a week of programs for schools should be televised during term time.

The broadcast of news and current-affairs programs is made mandatory on all companies by the IBA; other mandatory programs include two weekly plays, weekly documentaries, a weekend program relevant to the arts or sciences, and a Saturday afternoon sports program. Many program companies have achieved considerable success in selling their television programs overseas. In 1972-73, Associated Television sold programs worth nearly 11 million to over 100 countries, the majority going to North America. Sales by other ITV companies total several million pounds.

Fig. 13-4. A BBC TV cameraman filming a show jumping event.

REGIONAL AND LOCAL BROADCASTING

Regional broadcasting is recognized as an essential part of the public broadcasting service for two reasons: first, regional contributions add individuality and variety to the national networks and, secondly, regional and local broadcasting provides an opportunity for audiences in the regions to see or hear programs reflecting the special interests of their areas. A major contribution is being made by radio broadcasting in the form of 20 BBC local radio stations and the independent local radio stations being set up by the IBA.

British Broadcasting Corporation

The three BBC regions of Wales, Scotland and Northern Ireland produce a larger number of programs for purely local consumption than do the English regional television stations. There are no BBC local radio stations in these areas.

BBC Wales Radio is a bilingual service; there is an even balance between English- and Welsh-speaking programs in the locally produced output of some 30 hours a week. BBC Wales Television was established in 1964 and, in addition to its contributions to the BBC national network, produces some 12 hours a week of programs especially made for Wales, of which some seven hours are in Welsh. News, drama, light entertainment, music and school programs are among the programs especially made by BBC Wales. The BBC Welsh Orchestra is the only full-time professional symphony orchestra in Wales. The BBC commissions work directly from Welsh-born composers and playwrights.

Similar programs are produced by BBC Scotland, which also broadcasts regular news programs in Gaelic. Program output from BBC Scotland totals about 40 hours a week on radio and nine on television.

In Northern Ireland some six hours of television and 26 hours of BBC Radio Ulster are broadcast weekly from Belfast studios. Among programs for local viewers and listeners are news bulletins and series on sports and farming, local entertainment and documentaries, and educational series for Northern Ireland schools.

The policy and concern of the Scottish and Welsh services are controlled by the respective National Broadcasting Councils; Northern Ireland and the eight English television regions have advisory councils. Each region has considerable choice in the making and presentation of locally produced programs, but control is centralized in London for finance, building and senior staff appointments, London also provides such central services as audience research, scientific research and staff training schemes.

BBC Local Radio

BBC local radio had a considerable impact on the arts. For example, BBC Radio Manchester encouraged the setting up of a chamber orchestra which broadcasts regularly; a brass band championship is also organized by the station. BBC Radio Humberside has established a successful drama competition for local writers. The local stations between them carry many regular programs designed to stimulate interest in local music, theatre and other arts.

Although the stations are designed to cater to the needs of the local community, they supplement local material by broadcasting each day programs from the national networks (generally music from Radios 1 and 2 and national news from Radio 4). In turn, there are also benefits for the BBC national network as the latter has immediate access to local sources for important news developments. Each station has equipment connecting it directly with newsrooms in the appropriate television region, and in London a local radio unit has been set up to ensure a two-way flow of news between London and the local stations. The national radio networks occasionally transmit programs originally broadcast on a local station.

A local radio council appointed by the BBC advises each station. Finance comes from the television license fee. Grants are sometimes made by local authorities to assist in the making of educational programs.

INDEPENDENT BROADCASTING AUTHORITY

In addition to program services provided by the BBC, several are offered by the IBA.

Television

From the outset, independent television was a regional system, the authority dividing the country into a number of separate areas each served by an independent program company. The five largest companies, known as the "network companies," produce the bulk of nationally networked programs and also programs for viewers in their own areas. The remaining 10 companies, known as the regional companies, select network programs made by the other five but add a large number of local programs made by themselves for local viewers. The regional companies also produce programs which are shown in other independent television areas. About

9% of nationally networked programs originate from the smaller companies. In Wales there are some 5.5 hours a week in Welsh, and Channel Television, which serves the Channel Islands, broadcasts some news and weather forecasts in French. A regional news coverage, often in the form of magazine programs, is an important part of the output of 14 companies. (See Fig. 13-5.).

Of the 9300 hours of different programs shown over the independent television system as a whole in 1973-74, some 7500 hours were produced in the studios of the program companies. Some two-thirds of the 7500 hours were local programs, the remainder being nationally networked.

Radio

The IBA, as with independent television, builds, owns and operates the transmitters and the programs are produced by individual program companies appointed by the IBA. The latter has powers over the planning and quality of programs and the program companies are obliged to maintain a proper balance and wide range in their subject matter.

Each program company is appointed for an initial term of three years, and yearly extensions may be granted by the IBA at the end of the first term and each subsequent year, subject to satisfactory performance. Measures can be taken to prevent an accumulation of interests by an individual or corporate body in more than one sound program contract, and there are also provisions to protect local newspapers which are, in most

Fig. 13-5. A specially adapted Range Rover serves as an outside broadcasting unit for Independent TV news. Its compact size enables it to reach newsworthy events quickly.

cases, offered a shareholding in the company, although no newspaper with a monopoly in an area is permitted to acquire a controlling interest.

The main purpose of the stations is to provide a radio service for the local community. The IBA is required to appoint local advisory committees to cover all the areas where there are radio stations whose function is to give the authority advice reflecting the range of tastes and interests of people living in the area. Local authorities nominate one-third of the members and no member is permitted to have any advertising interests. In addition to advising the IBA, the committees provide the IBA with a channel for local reactions to the output of the individual stations, and an indication of whether a station is providing a worthwhile service to the community in its area.

The new radio stations are financed by advertising on the same principles as independent television; time is sold by the companies to advertisers, but the latter are not permitted to sponser any programs.

The local radio stations play an important part in community life by broadcasting local news and weather reports and covering local events of interest to members of their audience. Most stations also include items on sports and religion, while some have short stories, children's programs, phone-in programs in which local listeners can participate, and competitions. Some of the new stations broadcast for 24 hours a day.

Local Community Cable Television Experiments

In 1972, the government accepted the first application of a cable television company to mount local community television experiments. Greenwich Cablevision Limited, serving some 15,000 homes in the Woolwich area of south London, began transmissions of the local community television service in July, 1972. Four other cable television companies had their applications accepted in September, 1972; Rediffusion Limited at Bristol serving some 20,000 homes, British Relay at Sheffield (30,000); Radio Rentals/EMI Limited at Swindon (12,000); and Wellingborough Cablevision Limited at Wellingborough (5000). In addition, in March 1975, the home secretary authorised in principle a further local cable television experiment in Milton Keynes over a post office Network. (See Fig. 13-6).

The licenses (given under the Post Office Act 1969) authorized the services provided by broadcast relay organizations, covered the period to July, 1976, but in order to allow the committee of inquiry into the future of broadcasting time to complete its report, the licenses were extended for another three years.

The aim of the experiments is to provide a purely local community television service without extra charge (at least for a period) to cable television subscribers. Their purpose is to discover whether specifically local programs covering a small geographical area can in time meet a need for the viewers and be financially viable. The programs are usually transmitted in the early evening and subjects covered include local activities and events and local current affairs. General entertainment programs are not permitted.

Originally it was intended that the companies should finance their programs out of normal subscriptions received from viewers for reception of BBC and IBA programs relayed over the wired system. By early 1975, some of the companies faced financial difficulties and one reduced its broadcasts. The home secretary subsequently agreed to allow advertising to be included in their transmissions on the basis that similar standards apply to the content of advertisements as those which operate on independent television. Not all of the stations decided to use advertising, and two, Bristol and Wellingborough, closed down.

The Cable Television Association of Great Britain looks after the interests of radio and television cable services. It has some 50 members, all of which are companies holding home office licenses to operate cable distribution systems. It also has about 30 associate members who have interests in the industry, such as the manufacture of equipment.

TREATMENT OF CONTROVERSIAL SUBJECTS

When broadcasting first began, all controversial subjects were strictly excluded from the programs in deference to the susceptibilities of the listening public which, it was feared, might be offended by them. It soon became clear that this ban was detracting from the value of the service, although it was agreed to be of the utmost importance that the impartiality of the BBC should be above question. Following a recommendation of the Crawford Committee, the ban was lifted in 1928, and the BBC was free for broadcasting on controversial matters provided that it excluded any expression of its own opinions on current affairs or matters of public policy, and observed due impartiality.

Political Broadcasting and Current Affairs

The importance of radio as a medium for spreading political ideas and knowledge among a widening public was immediately recognized by the

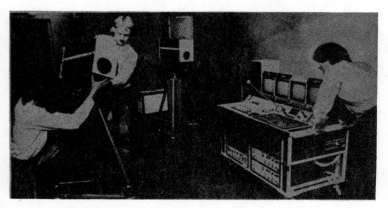

Fig. 13-6. Local community TV experiments: the production studio of Sheffield Cable-vision.

political parties, who seized upon the opportunity. Although at first it proved difficult for the BBC to secure agreement among them in the arrangement of balanced broadcasts, considerable success had been achieved by 1935 and the recommendations of the Ullswater Committee endorsed the practice of the BBC as it had been built up.

Broadcasting on political issues was virtually suspended for the duration of the Second World War, but never ceased entirely. Moreover, parliamentary debates were always fully reported in the daily news bulletins. There were frequent broadcasts by government spokesmen which kept the nation informed of the conduct and the progress of the war; and the series, *The Week in Westminster* (which is an impartial radio account of the week's proceedings in parliament), continued uninterrupted throughout the war years. In 1945, a daily factual and impartial account of proceedings in parliament was introduced known as *Today in Parliament*, and in 1946 the broadcasting of this item became a formal obligation: it now takes place every evening on Radio 4 when parliament is in session.

On television there is a weekly report on parliamentary affairs on BBC 2 and the news programs of BBC and Independent Television News (ITN) present regular reports on important debates in their daily bulletins. By act of parliament the IBA is required to ensure that all news is presented with due accuracy and impartiality.

The question of broadcasting parliamentary proceedings has been discussed on several occasions. In 1966, a House of Commons Select Committee recommended that a two-month experiment in the televising of commons proceedings on closed circuit should take place. The recommendation was debated in November, 1966, but the motion supporting the experiment was defeated. Experimental closed-circuit television and radio broadcasts of proceedings in the House of Lords took place in 1968, however, and a 4-week radio experiment was carried out in the House of Commons in the same year.

In October 1972, the House of Commons defeated a motion to approve an experimental period of public broadcasting by radio and television of proceedings of the House. In 1975, the House of Commons again debated the issue of parliamentary broadcasts and voted in favor of an experimental 4-week period of public sound broadcasting. At the same time a motion to authorize public television broadcasts for an experimental period was defeated. Following the results of the experiment, members of parliament will decide whether they are in favor of the permanent public broadcasting of their proceedings.

The rules governing post-war party political broadcasting were originally established by an arrangement reached in 1947 between the government, the parliamentary opposition and the BBC. This provided that (a) in view of its responsibilities for the care of the nation, the government should have the right to use radio and television from time to time for ministerial broadcasts; and (b) time for controversial party political broadcasts should be allocated to the main political parties broadly in proportion to their vote at the previous general election.

In 1969, a revised agreement was reached concerning ministerial broadcasts. There are now two categories of ministerial broadcast. The first relates to ministers wishing to explain legislation or administrative policies approved by parliament, or to seek the cooperation of the public in matters where there is a general consensus of opinion. There is no right of reply by the opposition.

The second category of ministerial broadcast relates to more important occasions when the prime minister or a minister wishes to broadcast to the nation in order to provide information or explaination of events of prime national or international importance, or seek the cooperation of the public in connection with such events. There is a right of reply by the opposition to the second category of ministerial broadcast. The BBC provides facilities for both types of broadcast. In the case of the second category, both government and opposition present their viewpoints and the BBC and Independent Television arrange additional broadcast discussions of the issues between a member of the cabinet, a senior member of the opposition, and a representative of the liberal party. Ministerial broadcasts are normally transmitted by both the BBC and the IBA.

A similar form of political broadcast takes place after the chancellor of the exchequer has presented his main budget to parliament. Under long-established practice, the BBC invites the chancellor and a spokesman nominated by the opposition to broadcast on successive evenings during the week of the Budget debate. The broadcasts are transmitted by both television broadcasting organizations and by the BBC on radio.

Religious Broadcasting

Religious broadcasting has always formed part of the services offered by the BBC and by independent television. Programs include religious services, prayers, hymns, stories of Christian faith, discussions, drama, music and programs especially for young people. An average of three hours a week of religious programs are shown on BBC television and eight hours a week of program time on BBC radio are devoted to religious topics. There are also religious broadcasts on the BBC's regional services including programs in the Welsh language for viewers in Wales. There is at least one religious service broadcast for BBC radio listeners every day of the year. Two and a half to three hours of religious programs are shown every week on independent television. These include morning services on Sunday and an hour and a half devoted to religious subjects on Sunday evenings. In addition, the regional companies contribute weekday material of local interest.

When broadcasting first began, no items of religious controversy appeared. At present religious broadcasting, while reflecting the faith of Christians and exploring the relevance of Christianity in the life and work of the churches, is, in part, devoted to the discussion of major moral, social and philosophical issues; in many cases agnostics, atheists and members of other faiths participate in these discussions with Christians. On occasions, talks are given by members of other religions. It is the responsibility of the broadcasting authorities to maintain a broad denominational balance in the general output of religious programs.

The central religious advisory committee, whose members include representatives of all the main Christian Churches in Britain, together with laymen, considers the policy of and receives reports on religious broadcasting by the BBC and IBA. The BBC has three regional advisory committees (for Scotland, Wales, and Northern Ireland) and the IBA is assisted by a panel of six religious advisors, representing the Church of England, the Roman Catholic Church, the Free Churches and the Churches in Scotland, Wales and Northern Ireland.

EDUCATION PROGRAMS

Broadcasting serves the educational interests and needs of all age groups, from children to adults.

Broadcasting for Schools

In 1924, the British Broadcasting Company started a series of experimental sound broadcasts for schools. These were continued after the company became the corporation in 1927, and in the same year schools programs were established on a permanent basis under the sponsorship and guidance of a body, now called the school broadcasting council for the United Kingdom, whose members include representatives from the department of education and science, local authorities, teachers' associations and other educational organizations. There are separate councils for Scotland and Wales and a committee for Northern Ireland.

The councils are responsible for determining general policy for school broadcasting on television and radio and the scope and purpose of each series. Their permanent staff includes a team of 22 education officers in various parts of the country who visit schools, meet teachers, and attempt to assess classroom response to the broadcasts and the educational effectiveness of the programs. Some officers are subject specialists. Continuing contact between producers of educational programs and children and their teachers enables program series to be more closely integrated into schools' curricula.

The planning and the execution of the broadcasts are the responsibility of the BBC school broadcasting departments in television and radio. Most series are accompanied by teachers' notes and many by illustrated pamphlets for the pupils. Each year the BBC produces for schools some 11 million copies of publications of various kinds including printed booklets and filmstrips.

The IBA is advised by an educational advisory council and two separate committees, one dealing with school broadcasting and the other with adult education. The production teams of the program companies are advised by special committees, composed of people who keep in close contact with schools and other educational interests in their regions. Educational liaison officers are employed by the companies. The plans of

254

all the companies producing educational programs are coordinated centrally with the IBA. Regular consultations take place with the BBC to minimize duplication of provision.

BBC radio broadcasts to schools cover most subjects of the curriculum, including English and reading, modern languages, music, drama, history, science and religion. Most of these broadcasts are designed for Britain as a whole, but some are produced only for schools in Scotland, Wales or Northern Ireland; in 1974-75 there were 94 series for Britain, nine for Scotland, 13 for Wales, and four for Northern Ireland.

In addition to using broadcasts directly, schools, particularly secondary schools, make more extensive use of them as resource material by tape recording radio programs, by video recording television, or by hiring certain television school programs made available on film. Tape/slide presentations are available to supplement both television and radio broadcasts, and a variety of teachers' and pupils' booklets, posters, and other materials are published in support of broadcasts.

It is estimated that almost all schools in Britain can receive BBC radio broadcasts. School broadcasts are also specially recorded for the BBC transcription service and are available to broadcasting organizations overseas.

Since 1957, both independent television and the BBC have provided nationally networked television broadcasts for schools. The programs are now watched regularly by children in over 28,000 primary and secondary schools throughout Britain. Programs are usually repeated to facilitate their inclusion into the school timetable. The BBC provides special programs for schools in Scotland, Wales and Northern Ireland, while the independent television companies produce a number of programs designed for their regions.

In 1974-75, the BBC broadcast 30 series for Britain and eight for Scotland, Wales or Northern Ireland. Independent television transmitted 47 series, of which 32 were nationally networked, and 15 shown in the regions. During 1972-73 the BBC and independent television took part in a joint project designed to improve the use of educational broadcasts in schools. Teachers at 106 schools made assessments of the effects on teaching and learning of school broadcasts.

A wide range of television programs is now available to schools, including subjects such as science, mathematics, modern languages, social history, geography, sociology, archaeology, music, drama and careers. There are special programs produced for preschool children and for backward children. (See Fig. 13-7.)

The IBA fellowship scheme provides for educationalists to study any aspect of the relationship between television and education at universities and other institutes of higher education in Britain.

Adult Education

Education programs for adults form an important part of the output of both the BBC and the IBA, the government having agreed in 1963 to extend the hours of permitted broadcasting for the inclusion of programs designed to give the viewer a progressive mastery of some skill or field of learning, whether vocational or recreational. These television broadcasts include series on science and language, art and literature and interests such as sailing and motoring. Many of the programs are supported by textbooks and pamphlets. The BBC is advised on adult education by the further education advisory council and the IBA by its adult education committee. Beginning October 1975, the BBC broadcast, for three years, weekly programs designed to help adult illiterates in reading and writing.

Among series provided by the BBC in 1973-74 were programs on social work, trade unionism, women at work, the European Community, television and society, crafts and poetry. There were also French and German language courses and special in-service education programs for teachers. Language courses are usually accompanied by books and gramophone records.

A variety of educational series were broadcast by independent television. Subjects covered included child development, famous historical places, the development of industry, farming, town planning, musical skills and appreciation, skiing and sailing, and advanced car driving. Independent television also previews some of its schools programs so that parents are able to see them and thus learn about current developments in education.

BBC radio began broadcasting adult education programs in the 1920s, but considerable expansion in the scope and variety of programs has taken place since the Second World War. A full schedule of programs is broadcast early on weekday evenings during an hourly session and selected series are rebroadcast on weekends. In 1973-74, BBC radio broadcast programs in German, French, Russian and Italian, and series on education, family and community, music and the arts, communication skills, contemporary affairs and leisure, which included a series providing an introduction to three countries: Greece, Portugal, and Yugoslavia.

The BBC also transmits early morning weekly programs in Asian languages on radio and television for immigrants in Britain. One of the main elements of the programs is English teaching.

The independent local radio stations include in their output a variety of educational broadcasts, usually involving local experts.

The Open University

Television and radio programs made especially for students of the Open University are broadcast twice weekly by the BBC at peak viewing and listening times. The main teaching methods of the university are a combination of television and radio broadcasts, correspondence work and summer schools. The aim of the university is to provide the opportunity of obtaining a degree comparable in standard to degrees awarded at other British universities to adults who can undertake systematic part-time study.

The television and radio broadcasts form one of the most important teaching elements. Members of the BBC staff are full-time members of the

Fig. 13-7. A scene from a London Weekend TV drama series *Affairs of the Heart*, based on stories by Henry James.

university's faculty course teams which prepare the teaching material. The BBC is also formally represented on the university council and, by invitation, on the senate. In 1975, the BBC's Open University productions department produced some 843 hours of radio and 962 hours of television in support of 84 courses and part courses prepared by the university.

Children and Television

Programs designed especially for young children are televised daily. Subjects include stories, cartoons, music, quizzes, sport, animals and comedy. The most popular BBC program is *Blue Peter*, a general information program which is shown twice weekly, while another weekly series *Search*, enables children to discuss serious contemporary issues. A regular part of independent television output is drama and adventure stories; magazine and information programs, and entertainment and general knowledge series, are also broadcast. Thames Television produces a popular series called *Magpie*, which is a magazine program for children up to 12 and which encourages active participation on the part of its audience. Associated Television has a program *Today is Saturday*, which introduces a variety of material in terms that children understand.

The impact of television on young people is the subject of considerable discussion, not least by the broadcasting authorities themselves who are aware of the problems faced by program producers when, for example, they make drama and adventure series in which some violence occurs. It is felt that children are more vulnerable than other sections of the population when exposed to violence on the screen. Both main channels transmit programs designed for general family viewing in the first half of the evening and steps are taken to ensure that material unsuitable for young children is not screened at this time.

Under the 1973 Independent Broadcasting Authority Act the IBA is required to draw up a code giving guidance about the showing of violence when large numbers of children may be watching. The act requires the IBA to exclude program material "likely to encourage or incite to crime or to lead to disorder or to be offensive to public feeling." The BBC also operates a code designed to limit the amount of violence shown on television.

Promotion of the Arts

The BBC exerts a powerful influence on the musical life of Britain. It has 12 orchestras (including those playing light music) and employs about a third of the full-time professional musicians in the country. There is also a professional choir of 28 singers, the BBC singers.

Among the orchestras is the Academy of the BBC, formerly the BBC Training Orchestra, formed in 1966 for the purpose of training qualified young musicians (aged between 18 and 26) and of providing extensive orchestral experience immediately following an instrumentalist course at a school of music; the orchestra broadcasts every week on Radio 3 and gives up to 12 public concerts a year.

One of the most important BBC musical projects is the annual 8-week season of Henry Wood Promenade Concerts which have been organized by the BBC since 1927. The BBC Symphony Orchestra plays at many of the concerts, but other British and overseas orchestras take part; in the 1974 season some 20 orchestras and ensembles participated, including orchestras from the United States and Australia. Some 34 conductors took part in the 55 concerts. The scope of the concerts has widened considerably in the past ten years, with the series including works by modern composers as well as the traditional classics; the audience on the BBC home services and the BBC world service amounted to many millions.

The BBC also contributes to the success of music festivals in Britain by regularly broadcasting concerts from the Edinburgh, Aldeburgh, Cheltenham and other festivals. Works are occasionally directly commissioned by the BBC, for example, Benjamin Britten's opera, *Owen Wingrave*, which was written especially for television.

The BBC makes a substantial contribution to drama in Britain. It has its own drama repertory company, and an English by radio repertory company. Many drama works are specially commissioned by the BBC. A considerable sum of money is spent by the BBC each year as copyright payments to authors and composers.

Many independent television companies make grants to regional arts associations, and to arts festivals. Yorkshire Television helped to establish the Yorkshire regional arts association; it also assists the association in a scheme sponsoring a string quartet based in the music department of the University of York. Westward Television sponsors an annual art exibition, while HTV (Harlech Television) has grant-aided the South West Arts Association and assisted the Bath Festival, the Bournemouth Symphony Orchestra and the Bournemouth Sinfonietta. The two companies in London, Thames and London Weekend, contribute to the Greater London Arts Association. Contributions to repertory theatres, art galleries and arts festivals in the north of England are made by Granada Television. Scottish Television administers an annual Theatre Awards Scheme designed to encourage drama in Scotland.

Independent television also covers important arts events such as the Edinburgh Festival, the National Eisteddfod in Wales and Glyndebourne Opera.

ADVERTISING

Under the terms of its license, the BBC may not without consent of the home secretary broadcast any commercial advertisement or sponsored program. The policy of the BBC, therefore, is to avoid giving any publicity to any firm or organized interest, except when this is necessary in providing effective and informative programs.

The broadcasting of advertisements during IBA transmissions is regulated by the Independent Broadcasting Authority Act 1973, which prohibits the sponsoring of programs by advertisers but allows the program companies to sell time for advertising. Advertisements may be inserted at the beginning or end of programs or during the natural breaks in the programs. They must be clearly distinguishable as such and be recognizably separate from the programs. The time given to them must not be so great as to detract from the value of the programs as a medium of information, education and entertainment. The amount of advertising on television is limited by the IBA to an average of six minutes an hour with a normal maximum of seven minutes in any clock hour (for example, 2 p.m. to 3 p.m.).

It is the duty of the IBA to draw up, and from time to time review, a code governing standards and practice in advertising and prescribing the advertisements and methods which are prohibited. The code is drawn up by the IBA in consultation with its advertising advisory committee and its medical advisory panel, and the home secretary.

The advertising advisory committee has nine members under an independent chairman and advises the IBA on the control of advertisements with a view to the exclusion of misleading or unsuitable advertisements; three of the members (all of whom are at present, women) are representative of the public as consumers, four are from advertising interests connected with the standards of advertising in all media and the remaining two represent professional medical associations. The medical advisory panel advises on medical and allied advertisements.

Advertisements are examined by the IBA's advertising control division and by a specialist advertising copy clearance set up by the program companies under the aegis of the Independent Television

Companies Association. Some 8000 scripts are dealt with every year, of which about 25% are returned for amendment because they contravene the code. Advertisements are excluded from broadcasts to schools, religious programs and formal royal occasions. No advertisement of a religious or political nature is permitted.

The advertisements are mainly for mass consumer goods, such as food and drink, chocolates and sweets, tobacco, household soaps and detergents, paint, petrol, cosmetics and proprietary medicines. Local advertisers publicize events of local interest, employment vacancies and goods on sale in local stores. In 1965, the government prohibited the advertising of cigarettes on television in view of growing evidence of a link between smoking and lung cancer. No advertisements for betting are allowed; by custom, the distillers do not advertise on television.

All advertisements for independent local radio must also comply with the IBA Code of Advertising Standards and Practice and a procedure has been established to ensure they do. The amount of advertising is limited to nine minutes in any clock hour.

COPYRIGHT IN BROADCASTING

The law of copyright in Britain is embodied in the Copyright Act 1956. The act is broadly based on the recommendations of a committee appointed in 1948 to study the law of copyright in the light of modern developments, and it contains the following provisions which related to broadcasting for the first time:

1. It creates a performing right in radio and television broadcasts to enable the BBC and the IBA to control the public showing of their broadcasts to paying audiences.

2. It gives broadcasting organizations the right to (a) control the making of films from their television broadcasts, otherwise than for private purposes, (b) control the making of recordings from their radio broadcasts or from the sound part of their television broadcasts, otherwise than for private purposes, and (c) control the rebroadcasting of their radio and television programs (but not the relay of these programs by broadcast relay companies).

3. It establishes a performing rights tribunal which has jurisdiction to deal, among other things, with (a) disputes between the broadcasting organizations and collective bodies representing the owners of the performing rights in literary, dramatic or musical works, and (b) disputes between the broadcasting organizations and persons who have been refused a license to show television programs to a paying audience, or who claim that the terms of a proposed license are unreasonable. The tribunal has no jurisdiction where the performing rights are held by an individual copyright owner, or in respect of mechanical rights.

The act also contains a provision that fair dealing with a literary, dramatic or musical work shall not constitute an infringement of copyright if it is for the purpose of conveying news or current events to the public by means of broadcasting. It provides, moreover, that although the inclusion of an artistic work in a television broadcast without the consent of the copyright owner constitutes an infringement of copyright. Copyright is not infringed by the televising of (a) works of architecture, (b) sculptures or works of artistic craftsmanship permanently situated in public places or in premises open to the public, and (c) artistic works included in a television broadcast by way of background or otherwise only incidental to the principal matters represented in the broadcast. Copyright subsists for 50 years. An international convention, ratified by Britain in 1963, protects the rights of performers and broadcasting organizations.

A European Agreement for the Protection of Television Broadcasting, prepared by the Council of Europe, was signed by the British representative in 1960. This agreement gives to the various European television companies the same protection in Britain as is enjoyed by the BBC and the IBA, and grants reciprocal protection to BBC and IBA television broadcasts in European countries.

The BBC and IBA pay an annual lump sum to the Performing Rights Society, a central body which holds the performing rights in most copyright music, in return for a license to broadcast music controlled by the society in their programs. An annual fee is also payable for the right to broadcast commercial phonograph records for specified weekly periods (which are known as "needle time"). By agreement with the professional bodies concerned, fees are specified for the broadcasting of prose, poetry and dramatic works.

AUDIENCE RESEARCH

There are no means of estimating the exact sizes of audiences for radio and television broadcasting, although the numbers of current television receiving licenses give some indication of the total potential viewing audience. There is also no means of obtaining definite evidence, such as sales figures, of the reactions of the public to any particular broadcast. Although it is generally agreed (and inherent in the terms of the preamble to the BBC charter) that broadcasting should not be governed solely and automatically by what is thought to please the listeners or viewers, it is nevertheless obvious that it must conform as far as possible to public taste, since, without an audience, it would have no purpose at all. Some form of audience research, therefore is, essential.

The system of audience research conducted by the BBC began in 1936, after a careful study of the various methods that could be employed. The size of an audience is measured by interviewing some 2250 people, including children, daily in Britain and ascertaining which programs this representative sample of the population have listened to or viewed during the preceding 24 hours. Different people form the sample each day so that more than 800,000 people are contacted annually. A list of all programs broadcast nationally (and some transmitted in certain areas only) is compiled, against each of which figures are given, indicating the proportion of people found to have listened to or looked at the programs.

The opinions of audiences are obtained through listening and viewing panels, the volunteer members of which are selected of a representative cross-section of the public and are prepared to answer questions about the programs that they normally see or hear. The members of the panels (some 6000) regularly receive questionnaires about forthcoming broadcasts which seek frank expression of opinion and asks the panel to sum up their reactions using a simple scale. This leads to reaction indices for television programs and to general evaluations for radio programs. Longer questionnaires provide material for the production of program reports which try to give a balanced picture of the opinions expressed, bringing out the majority view and the various minority views.

Trends in public taste which cannot be discovered by an examination of the results of interviews and the reports of the listening panels are the concern of a section of the BBC's audience research department which carries out special investigations. Recent examples have included a study of the impact on viewers of violence in television programs, and another on public attitudes to the BBC's television and radio news services.

The Independent Broadcasting Authority Act 1973 requires the IBA to "ascertain the state of public opinion concerning the programs." The basic figures of the number of viewers of independent television programs are supplied each week and monthly by an independent research organization, Audits of Great Britain Limited, through the Joint Industry Committee for Television Advertising Research (JICTAR), which is responsible for the service. Automatic electronic meters are fixed to television sets in a representative sample of 2650 homes to record minute by minute the time during which the sets are switched on and the channels to which they are tuned. The sample audience provides further data by filling in diaries and giving personal details about themselves so that the composition of an audience can be discovered.

The main source of information concerning how much the audience appreciates or enjoys the programs which they choose to view is the IBA Audience Appreciation Service, which operates with the help of a representative panel of about 1000 adult viewers in the London area and random postal samples of about 2000 electors in each of the ITV regions outside London; each respondent provides information, recorded in a specially designed diary, on his or her enjoyment of the programs viewed. Processing of this data leads to an average score of "appreciation index" for each program which aids program planning. These measures are supplemented by regular and special opinion surveys, undertaken both by the IBA and the program companies on different aspects of television.

EUROVISION

As well as taking part in the exchange of radio programs arranged between the member countries of the EBU, the BBC is a regular contributor to the European television network. The exchange of television programs between Western European member countries is arranged by the EBU, which maintains an International Television Coordination Center (Eurovision) in Brussels. One of the features of Eurovision is the thrice daily news exchange: in 1974 the EBU relayed over 92,000 television news items through Eurovision as well as 660 sports programs.

Apart from the news exchanges, BBC television offered 67 programs and received 97. The BBC also allocated studios on 77 separate occasions to 14 television organizations. In the same year the BBC allocated 17 film crews to 10 different organizations (French television was the largest user) and in return received the use of 55 film crews from 11 organizations. In 1973, the BBC supplied film of the European Athletics Cup in Edinburgh to 18 countries through Eurovision.

The BBC also cooperates in joint production of programs with European broadcasters. Some independent television companies, notably the ATV Network Limited, also cooperate in joint productions with European broadcasting organizations.

INTERCONTINENTAL TELEVISION

Intercontinental television relays by means of communications satellites are well established. They began in 1962, when the first live program exchanges between the United States and Europe were made via Telstar. Since then they have increased in range and scale as synchronous satellites, which are stationary relative to the earth and thus permit continuous two-way transmission, have been commissioned.

Direct links were achieved in 1964 between Europe and Japan and in 1966 between Britain and Australia. In 1965, a consortium of states formed Intelsat which organizes broadcasting by means of satellites covering the world, reaching earth stations in all continents. The network has continued to expand and there are now some 88 earth stations in 64 countries.

Live broadcasts by the BBC and IBA by means of the satellite system have increased in importance. Recent examples include coverage of the 1972 Olympic Games, the United States presidential elections in 1972, and the war in the Middle East in 1973. The satellite system has also enabled world collaboration in news transmissions to increase, and contacts are being established between the news departments of many broadcasting organizations throughout the world.

Chapter 14
Broadcasting in Ireland

Louis McRedmond
Head of Information and
Publications
Radio Telefis Eireann

Shortly after the establishment of the Irish Free State in 1922, a public debate began on the provision of a national radio service.

ESTABLISHMENT OF BROADCAST SERVICE

Proposals for a commercially operated station were considered, but the government decided in favor of a service provided directly by the state. The Department (Ministry) of Posts and Telegraphs set up a 1-kW transmitter in Dublin and broadcasting commenced on 1 January, 1926, with an inaugural address by Dr. Douglas Hyde, founder of the Gaelic League and later President of Ireland.

The Dublin station was officially designated Dublin 2RN, a whimsey evoking the concept "Dublin-to-Erin." A similar station opened shortly afterwards in Cork, under the more prosaic label of Cork 6CK.

Both stations had an effective range of little more than 30 miles from their transmitters. A gradual extension of these reception areas came about through the development of improved radio sets, rather than from any increase in signal strength. National reception of the Irish radio service was not possible until a 60 kW (later increased to 100 kW) transmitter was installed near Athlone in 1932.

The program on 1 January 1926, consisted of 16 items, mainly musical, but including a weather report, and ran a little short of three hours. This level of output was maintained from the beginning, despite rigid financial stringency. The chronic shortage of money stimulated much inventiveness. The director of broadcasting gave song recitals himself, and outside broadcasts brought theatrical and musical groups to the notice of a wide audience. Childrens' programs were carried and it seems that the 2RN coverage of a hurling match in August, 1926, was the first live commentary on a field sport to be carried in Europe.

A MATURE SERVICE

Irish radio rapidly matured when radio accommodation was provided in the rebuilt general post office in Dublin and the Athlone transmitter made the service available throughout the country. Hours of broadcasting were expanded, the station orchestra grew into a full symphony orchestra, actors were grouped into the service's own repertory theatre company, and scripted talks gave way to free-range live discussions. A proper news department was set up and mobile recording studios made possible the origination of many programs in the provinces. Without legislative sanction, the service adopted for itself the call-signal "Radio-Eireann" (Radio of Ireland) and this became its accepted title, although on the dials of radio sets it was still identified as Radio Athlone.

It would be difficult to quantify the achievements of Irish radio in the 35 years before the arrival of television. At the practical level, it provided employment at home for many talents, artistic and scientific. It created a new industry as real as a large manufacturing enterprise but more profound in its implications.

Socially, Radio Eireann improved the quality of life in Ireland, especially in rural Ireland, out of all recognition and was an important unifying factor in the new state. Culturally, it introduced professional theatre, classical music and many other features of cosmopolitan living to people who would have had no access to them. At the same time, it brought about a renaissance of traditional arts, most of all music where it encouraged the revival of old forms in a modern idiom. It helped make the Irish language a medium of modern communication and, through the Radio Eireann Players, it evoked the principle of drama-for-broadcasting as distinct from the mere transmission of stage productions. Its success in this field was acknowledged twice in the 1960s by the award of the prestigious Prix Italia to drama productions by the Irish Radio Service.

TELEVISION ARRIVES

From the late 1940s, British television could be received along the east coast of Ireland. The question of providing an Irish service was considered by a government-appointed commission. Once again, the government decided in favor of public-service broadcasting rather than commercially motivated private enterprise.

The Broadcasting Authority Act, 1960, established a state company to be known as the Radio Eireann Authority, which was empowered to operate both the existing radio service and the new television service. This corporation was given autonomy over all programming, subject to certain powers reserved to the minister for posts and telegraphs or to the government. Its authority, or board of governors, was appointed for a term of years by the state. With minor changes, including a change of name in 1966 to Radio Telefis Eireann (Radio-Television of Ireland), this is the system of public service broadcasting which continues to operate in Ireland today.

TELEVISION PROGRAMMING TODAY

The studios for the television service were built at Donny-brook (see Fig. 14-1), a south city suburb of Dublin, on the grounds of Montrose House, an appropriate choice, since Montrose had been the family home of Annie Jameson, mother of Guglielmo Marconi, the inventor of radio.

Five main transmitters on mountain tops to carry the network to all parts of the country and the Irish television service came on the air on 31 December, 1961, with an inaugural address by Mr. Eamonn de Valera, President of Ireland.

For a small service by international standards, Irish television carried a high proportion of home-produced programs from the beginning. By the mid-1960s, these totalled more than 60% of output. The proportion has declined in recent years to some 48%, because the extension of daily hours of broadcasting has required the use of more imported material.

The home-produced content of the schedule covers every aspect of the television spectrum: news and current affairs, light entertainment (see Fig. 14-2), serious drama, sport documentaries, traditional and classical music, as well as special service subjects such as agriculture, education and religion (see Fig. 14-3). The "colorization" process, begun in 1970, has been completed and all output now is in color.

It might be argued that Irish television attempted too much too soon with inadequate resources, so that its programs were of variable quality.

Fig. 14-1. RTE television studios, Donnybrook, Dublin 4 (courtesy RTE).

266

As it settled down, it became more sophisticated and more professional. Its prize-winning record at international festivals in recent years has been formidable, especially with drama, educational, and, above all, agricultural programs. Some home production is in the Irish language, including one major news bulletin every night.

RTE's imported programs come in roughly equal proportions from the United States and the United Kingdom, with a small intake from Canada, Australia and continental Europe. RTE broadcasts most feature films and American television series before these appear on the British services. It even carries occasional British productions before they are seen in the United Kingdom. Such judicious buying and scheduling is vital to the Irish service which must compete, in much of its reception area, with the three British channels available off-air and, in large centers, on cable systems. In this area of multi-channel reception RTE holds almost 50% of total viewership.

In the south and west of Ireland, where the British services cannot be received, a sense of frustration developed in the early 1970s. Viewers began to demand the same multichannel reception enjoyed on the east coast and along the border with Northern Ireland. Their resentment at the lack of viewing choice was heightened by the public service obligation of RTE to meet the needs of majority interests as well as to satisfy a mass audience. In these circumstances, RTE's wide range of programming provided a disadvantage. Serious drama, educational programs and productions in the Irish language alienated a number of single-channel viewers who had no other service to which they could turn for alternative entertainment.

The government decided that a second service should be provided by RTE. Following a delay caused by the severe inflation of the mid-1970s,

Fig. 14-2. Light entertainment on Irish TV (courtesy RTE).

267

plans were initiated to establish RTE-2. Initially, this service will take some 80% of its programming from British and other foreign sources. The balance will consist of new home production. RTE-2 began operations November 2, 1978. (See Fig. 14-4).

The Irish are not passive viewers. To perhaps a unique degree, they see television as a public property about which they are entitled to express their opinions. This they do directly to RTE, in the newspapers and through their public representatives. Although the opinions expressed are often at variance with one another, some consistent trends can be noted: a critical awareness of public affairs programming, for example, and an anti-metropolitan bias in the provinces which reveals itself as suspicion of whatever comes out of Dublin. (See Fig. 14-5.) Such strongly voiced public attitudes condition and assist program planning in RTE. They also influence policy making, such as a recent experiment in limited regional television for the southwest of the country and the use of new mobile facilities for the origination of more network programs in the provinces.

RADIO SINCE TELEVISION

It took several years for radio to adapt to the challenge of television, not least because of the loss of skilled staff to the new medium. By the late 1960s, RTE radio had acquired new life through the extension of broadcasting hours. This facilitated the provision and development of popular interest programming in the morning and early afternoon, with high standard minority or specialist programming during the television hours of late afternoon and evening.

Popular music, consumer advice and magazine-type discussions predominated in the earlier part of the day. The evening was largely given over to the plays, classical and traditional music, in-depth discussion of the arts or politics, lecture series on historical subjects and similar programming in which Irish radio had developed particular finesse. New formats were given to sport and agricultural reporting. Frequently, news bulletins were broadcast and a significant innovation was a news magazine providing immediate analysis of the events of the day at home and abroad. This has remained broadly the pattern of programming on the national radio service.

There have been technical changes also. To improve the quality of reception, a VHF (FM) radio network was built in 1966. With replacement of Athlone by a high-power 500 kW medium-wave transmitter at Tullmore in 1975, this VHF system became available for alternative programming. A choice of radio programs is now provided on weekends and it is intended to expand the separate service on VHF to make it a second radio channel in its own right.

Local radio has been initiated with an hour-long program on weekdays, produced in Cork and broadcast on the Cork transmitter (which "opts-out" from the national medium-wave service for the purpose). There are studios not only in Cork but also in Galway, Limerick and Waterford to provide regional inserts to the national network. Community radio has been developed successfully with a mobile studio and transmitter unit which visits a variety of centers throughout the country.

Typically, some five hours of programs, compiled and presented by local people, are broadcast daily during a week's visit. Reception on VHF or medium wave is confined to a 5-mile radius of the mobile unit. The British services have shown much interest in this development and the BBC has announced its intention to provide a community radio in Wales on the Irish model.

In Dublin, a new radio center was completed in 1974 on the Donnybrooke site adjacent to the television headquarters. The radio center ranks among the most advanced buildings of its kind in Europe, with 13 studios, stereo facilities, elaborately designed acoustics, and accommodation for the concert orchestra and the Radio Eireann Players.

RADIO na GAELTACHTA

A radio service for the Irish-speaking parts of the country (the Gaeltacht) was inaugurated in 1972 by RTE at the request of the government. Since the Gaeltacht regions are mainly on the west coast, the studio center was built at Casla in Connemara, County Galway, with satellite studios in County Kerry and County Donegal. Radio na Gaeltacht staff are almost all natives of the Gaeltacht themselves. The service of news, current affairs and entertainment programming in Irish is broadcast nightly for 3½ hours. It can be heard on its own medium-wave frequencies on the west coast and nationally on VHF. Radio na Gaeltachta also has a following in Gaelic-speaking areas of Scotland.

Fig. 14-3. Photo taken during a live Saturday night "Late Late Show." Compere, Gay Byrne (front right) (courtesy RTE).

Fig. 14-4. Launch of second TV channel, RTE 2, on November 2, 1978 (courtesy RTE).

OUTSIDE THE REPUBLIC

RTE maintains its own radio and television studios in Belfast. It has access to similar facilities in London and Brussels, where it has permanent offices and a journalistic staff. RTE radio can be received throughout most of Britain and in parts of northern Europe.

MUSIC-MAKING

As well as making radio and television programs, RTE maintains two large orchestras, the symphony orchestra and the light orchestra. These orchestras contribute to the broadcasting output, and are highly important features of Irish musical life. The symphony orchestra is the only full-time professional orchestra in the state. It has toured abroad with considerable success.

RTE has a permanent professional choir, the RTE singers. It also maintains its own professional theatre group, still known as the REP (Radio Eireann Players). The phonograph library holds the largest collection of records in the country.

PUBLICATIONS

RTE publishes a weekly journal, the *RTE Guide*, which provides radio and television program details (including those of Radio na Gaeltachta) with background information on selected programs. RTE also published occasional anthologies of broadcast material in book form.

270

AUDIENCE RESEARCH

RTE has its own audience research unit and also participates in multi-media research with the national and provincial press, the cinemas and the advertisers.

TAM (Television Audience Measurement), a service operated by an independent company, provides RTE with statistics on television viewing. All of these surveys and reports are a useful guide to the program planners in their efforts to cater for the different tastes and interests among the national audience.

EUROPEAN LINK

RTE takes about 20 film stories a week from Eurovision News Exchange, which links RTE daily with national television services all over the world.

The international film agencies also distribute their pictures through the Eurovision Exchange, and offer a broad spectrum of world coverage of important news stories.

RTE is an active member of the European Broadcasting Union.

CONSULTANCY SERVICE

RTE provides assistance to foreign broadcasting services in the design and installation of studio centers and transmission networks. The wide experience gained over 50 years of broadcasting is made available through the RTE Consultancy Service.

Fig. 14-5. A scene from "Famine" (courtesy RTE).

RTE has over 800 staff employed daily in the design, installation, maintenance and operation of radio and television systems. In addition to the personnel normally assigned to a consultancy project, members of this experienced staff are released to provide specialist service.

When necessary, key personnel from the organization for which the consultancy is being provided are taken into RTE for on-job training and familiarization of staff in the maintenance and operation of systems and equipment.

In choosing professional broadcasting equipment, the world is the market place. Because there are no manufacturers of broadcasting equipment in the Republic of Ireland, RTE can recommend the best, without bias. RTE's own installations are comprised of equipment from many countries, including Germany, United Kingdom, France, USA, Japan and Italy. Equipment is chosen for a project and recommended on the basis of suitability and cost, irrespective of country of origin.

Consultancy projects are financed by the recipient countries or by national and international development agencies.

FINANCE

Broadcasting has been financed by license fees and sales of advertising time since 1926. The separate radio license was abolished in 1972 and only a television license fee is now charged. The government determines the level of the license fee, including the supplement for a color receiver, and also approves the advertising rates. In 1976, license fees produced 47% of total income and advertising costs produced 45 percent.

Capital costs for building and equipment are met mainly by repayable interest-bearing exchequer advances.

ADVERTISING

Apart from one hour of sponsored radio programs on weekdays, all advertising on radio and television consists of short spots. No more than 10% of radio broadcasting time may be given to advertising and on television no more than 7½ minutes of advertising may be broadcast in any hour. Every advertisement must conform to a strict code of standards established by RTE.

BROADCASTING OBLIGATIONS

Under the Broadcasting Authority Act, 1960, and subsequent amending legislation, RTE is required to be impartial and objective in its presentation of news and current events, have regard for the privacy of the individual and refrain from broadcasting any matter "which may reasonably be regarded as being likely to promote or incite crime, or as tending to undermine the authority of the state." The minister of posts and telegraphs may prohibit the broadcasting of matter which, in his opinion, would fall into the categories referred to in the above quotation. RTE maintains a set of guidelines for its staff to assist them in observing the obligations defined by law. A Broadcasting Complaints Commission, whose members are appointed by the government, may adjudicate certain classes of complaints relating to alleged infringement of its obligations by RTE.

ADVISORY COMMITTEES

The RTE Authority is empowered to appoint advisory committees to assist in the formation of policy. At present, there are three such committees. They are concerned respectively with the operation of Radio na Gaeltachta, Irish-language programming on the national radio and television services, and educational broadcasting.

ORGANIZATIONAL STRUCTURE

The chief executive of RTE is the director-general, appointed for a term of years by the authority with the consent of the minister of posts and telegraphs. Under the director-general, who is both general manager and editor in chief, there are divisions for radio programs, television programs, news, advertising sales, engineering, financial control and personnel.

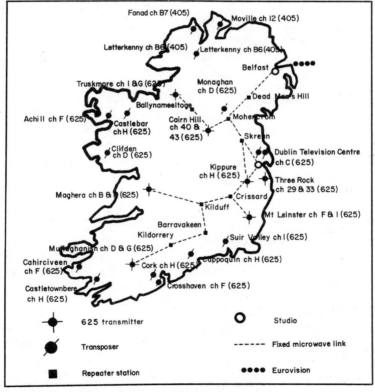

Fig. 14-6. Map of the national television Network.

273

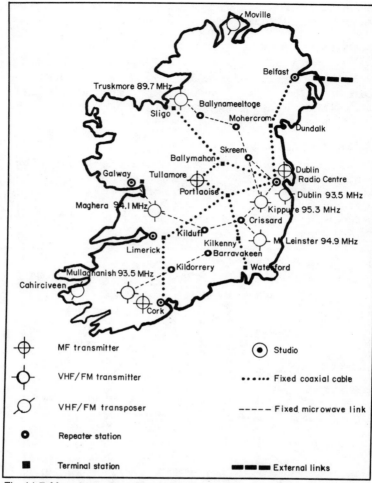

Fig. 14-7. Map of the national radio network.

The facilities locations of the national radio and television networks are shown on the maps in Figs. 14-6 and 14-7.

STATISTICS AND RELATED INFORMATION

Radio Receivers: Total in the Republic of Ireland: 949,000; homes with at least one set: 730,000 (93%); homes with more than one set; 180,000 (23%).

Television receivers: Total in the Republic of Ireland: 675,000, including 210,000 color sets.

Total staff: 1826.

Chapter 15
Broadcasting in Italy

Prepared by
RAI,
Italian Radio-Television

RAI—Italian Radio Television—is a shareholding company, based on total public participation, with legal and administrative centers in Rome. The share capital, amounting to 10 billion lire, is almost entirely held by IRI (Institute for Industrial Reconstruction).

THE INSTITUTIONAL SETUP

By an Act of Parliament (Law No. 103 of April 14, 1975, and following sentence No. 202 of the Constitutional Court of July 28, 1976) RAI is authorized to provide exclusively, on a national scale, the services of circular diffusion (broadcasting to all areas) of radio programs by air or by wire and television programs by air or by cable or any other medium. The concession given to RAI by the state to provide "this essential public service of outstanding general interest" (Article 43 of the Constitution, Reform Law No. 103 dated April 14, 1975) includes also the installation and technical operation of networks and systems intended for circular sound and television diffusion as well as the transmission by means of such systems, both within Italy and abroad, of a wide variety of programs. The current 6-year concession, which derives from the Convention of August 11, 1975, between the Ministry of Post and Telecommunications and RAI, necessarily establishes the status of RAI as a company of national interest (Article 2461 of the Civil Code).

The fundamental principles governing the public radio and television services are independence of viewpoint, objectivity of reporting, and sensitivity to variations of political and cultural expression, while, at the same time, respecting constitutionally guaranteed civil liberties. The aims of the services are to broaden citizen's participation in public life and to contribute to Italy's social and cultural development.

In providing the services for which it has the concession, RAI is bound to remain within the scope of the reform law and, where applicable, to follow any legislative ruling regarding telecommunications. It is also bound to respect any agreements reached by competent international organizations and to comply with any technical norms laid down by them.

The present institutional basis of RAI is parliamentary control and total public participation in the company.

HISTORICAL DEVELOPMENT

The origins of RAI date to 1924, when URI (Italian Radio Union) was set up, its share of capital being divided between RADIOFONO (Italian Company for Circular Radio Communications) which was the majority shareholder, and SIRAC (Italian Company for Circular Radio Listening). In the same year, on October 6, the Rome station of URI began the first Italian radio broadcasting service. Thereafter, by means of an exclusive 6-year concession, the state entrusted the provision of circular radio listening services to URI. Three years later, by Royal Decree (Royal Decree No. 2207 of November 17, 1929), URI was changed to EIAR (Italian Radio Listening Corporation), and until December, 1952, it was EIAR, which received the state concession to provide circular radio listening services. In 1931, EIAR was indirectly controlled by SIP (Hydroelectric Company of Piedmont): in 1933, SIP gained the absolute majority shareholding in EIAR, and in 1944, the name EIAR was changed to RAI—Radio Listening Italy.

At the convention of January 26, 1952, it was decided that the absolute majority of RAI shares be transferred to IRI. Thus, on the following March 30, SIP transferred to IRI a portion of its shareholding in RAI, amounting to 75.45% of the total share capital of RAI.

On April 10, 1954, as a consequence of expanding the television services, the company modified its name to RAI—Italian Radio Television. Ten years later, on December 21, 1964, SIP transferred its residual RAI shareholdings to STET (Telephone Shareholding Company). Finally, following the recent reform complete public participation in RAI was achieved with the transfer of the last privately held shares to IRI, giving it its present shareholding of 99.55% in RAI as against the 0.45% shareholding of SIAE.

RAI ORGANIZATION

The statutory organs are as follows:

1. The general assembly of shareholders
2. The Board of Directors which appoints
   a. the chairman
   b. vice chairman/chairmen
   c. the director general
3. Auditors

The board of directors consists of 16 members, of whom six are elected by the general assembly and ten by the parliamentary commission for the general orientation and surveillance of radio and television programs. Four of the members elected by the parliamentary commission are chosen on the basis of the recommendations of the regional councils. The term of office of members of the board is three years, and board members are eligible for re-election at the end of their term. Resolutions are adopted by a majority of votes, the chairman having the casting vote in the event of parity. The board is responsible for all aspects of the running of the company, except those reserved by law to the shareholders assembly.

The vice chairman/chairman is appointed from the membership of the board of directors. The board prescribes his functions.

The chairman is appointed by the board from its membership. He is entrusted with the company's legal representation and is answerable to his fellow board members. It is the chairman's responsibility to ensure that the company's management is achieving its objectives as well as ensuring that the company functions in accordance with the orientations of the parliamentary commission.

The director general, who is appointed by the board, sits at board meetings as a non-voting member. As executive director it is encumbent on him to ensure that the company provides a radio and television service, to ensure that board decisions are implemented, and that these are in accordance with the orientations of the parliamentary commission.

There are five regular auditors and two alternatives. Two plus one are appointed by the parliamentary commission and three plus one are appointed by the shareholders assembly. The auditors ensure control of the company's management, in accordance with Articles 2403 et seq. of the Civil Code.

RAI ORGANIZATIONAL STRUCTURE

The organization of RAI is prescribed by the Reform Law. The 1975 convention, along with relevant regulations, requires that the concessionary company be organized so as to ensure, above all, that the fundamental principles of the company, i.e., independence of viewpoint, objectivity of reporting, and sensitivity to a plurality of political and cultural expressions, are observed.

Operationally, this implies that productive activity should be undertaken with an eye to a balanced development of the company's productive capacity, and that the management of the services reflects the importance of the multiplicity of opinions expressed in Italian society. To this end the company is relatively decentralized with regard to both its administration and its production. Such decentralization ensures that RAI has an effective relationship with the differentiated reality of Italy and safeguards the impartiality and professionality of the information services.

The first stages of the company's restructuring, following the institutional changes in 1975, were reached in the 1976 fiscal year. In the process of reordering, priority was given, in the spirit of the reform law, to the activity of production.

The most significant events in the reordering process are as follows:

● January 30, 1976, commencement of the new *TV newsreels* and *Radio News Bulletins*

● April 14, 1976, the division of each of the television and radio networks into program planning structures, which, by means of a distribution of competences, takes account of technical factors and differences in listening and viewing audiences in the various daily timebelts throughout the weeks, covering the range of cultural and entertainment program

● July 5, 1976, new regional bases were established

● October 25, 1976, definition of RAI is divided into 18 central structures with 21 peripheral structures, linked by certain common functions

The central organization comprises first and foremost the structures for the indeation and realization of programs. These include: First Television Network (TV 1), Second Television Network (TV 2), First Radio Network (RF 1), TV Newsreel 1 (TG 1), TV Newsreel 2 (TG 2), Radio News Bulletin 1 (GR 1), Radio News Bulletin 2 (GR 2), Radio News Bulletin 3 (GR 3), Schooling and Education Programs Department (DSE), Administration of Stands and Access (DTR), Administration of Journalistic Services and Programs for Abroad (DE). Table 15-1 lists the hours of broadcasting devoted to each service and Table 15-2 the staff employed. The administrative services are ensured by the following supporting structures: Staff management (DP), technical management (DT), administrative management (DA), commercial management (DC). In addition, a secretariat of the board of directors (SC) has been set up, although certain functions in common between structures come under the direct control of the general manager; for example, coordination and planning, secretariat services, relations with abroad, *Premio Italia* and cultural and artistic events, program research and experimentation, legal affairs, documentation and studies.

Coordination of the television and radio networks, and of their respective supports, is entrusted to three vice directors general, each responsible for one part. The peripheral organization is divided into 21 regional bases. Associated with the regional bases of Milan, Naples, and Turin is a radio-television production center. Associated with the home base is a radio production center and a television production center. As a general rule, the centers produce a whole range of radio and television programs for the national networks and cover in their various stages those programs intended for local diffusion.

In the sector of journalism, services for both national radio and television and local radio and television are produced by the regional bases. In addition, regional bases contribute to the production of a variety of programs intended for the networks. Some notable examples of this are the Bolzano Base, which produces radio and television programs in German and radio programs in Latin for the minorities of Trentino-Upper Adige; the Trieste Base which produces radio programs for Friuli-Venezia Guilia in Slovene, and the Aosta Base which produces radio programs in French for the Aosta Valley.

The parliamentary services and the sports services, which are coupled on the basis of similar functions to the information services for the presidency of the republic and for the constitutional court, the offices of foreign correspondence, and the journalistic documentation services, are

278

Table 15-1. Radio and Television Program Broadcasting Hours in Each Program Category.

	Television				Radio			
	1974	1975	1976	1977	1974	1975	1976	1977
National Networks								
Network 1	3470	3265	3318	3276	6302	6387	6391	6380
Network 2	1782	1855	2633	3262	6046	6122	6122	5956
Network 3	—	—	—	—	5427	5300	5897	6371
	5252	5120	5951	6538	17,775	17,809	18,410	18,707
Local Networks								
in Italian	187	163	168	166	8245	8437	8764	12164
in German	514	543	564	552	4036	4083	4092	4136
in Ladin	—	—	—	—	150	151	158	155
in Solvenian	—	—	—	—	4432	4488	4500	4508
	701	706	732	718	16,863	17,159	17,514	20,963
Programming for abroad								
in Italian	—	—	—	—	5642	5552	5576	5575
in Foreign languages	—	—	—	—	5778	5793	5834	5818
TOTAL	5953	5826	6683	7256	46,058	46,313	47,334	51,063

further examples of unitarian structures operating for the journalistic sectors.

CONTROL OF RAI

In Italy, from the outset, the radio and television service assumed the character of a public service, understood as an organ of the state aimed at meeting certain general needs of the individual citizen and of the society. This concept of public service was defined in the Postal Code of 1936 (Royal Decree No. 645 of February 27, 1936) and the Postal Code of 1973 (Decree of the President of the Republic No. 156 of March 29, 1973), Article 1 of which establishes telecommunications as belonging exclusively within the domain of the state, and which in subsequent articles, regulates their operation and forms of concession. Reform law No. 103 of April, 1975, has reaffirmed that telecommunications in Italy fall within the domain of the state. RAI, as the concessionary of a public service, has had its activities subjected to a series of limits and controls aimed at guaranteed service consistent with the objectives of general interest.

THE PARLIAMENTARY COMMISSION

In order to better ensure the public scope of the service, the reform Law tried to establish at the various levels of management a privileged political relationship with parliament. The supreme organ of control of RAI is the parliamentary commission for the general orientation and surveillance of the radio and television services, consisting of 10 members selected jointly by the presidents of the two chambers of parliament from the representatives of the parliamentary groups. Besides electing the 10 members of the board of directors (outlined above) the parliamentary commission undertakes a vast activity of orientation and control over the concessionary. In particular, it:

● formulates the general orientations and principles to be observed in drawing up programs and controls their implementation, ensuring that they accord with the principles of public service

● lays down the norms for access to the radio and television media according to procedural patterns established by a permanent subcommission (regulations for access were approved by the commission on April 30, 1976)

● regulates the programs, *Political Tribune, Electrical Tribune, Trade Union Tribune* and *Press Tribune* directly

● indicates the general criteria to be followed when drawing up annual and pluriannual plans of expenditure and investments

● approves the overall plans for annual and pluriannual program planning, ensuring that these are implemented

● formulates the general criteria relating to publicity messages; it analyzes the content of broadcasts and telecasts and produces data concerning audiences and program popularity ratings.

PROGRAM PLANNING CONSTRAINTS

Special obligations affect overall program planning (Article 9 of the Convention of August 11, 1975). The duration of diffusion must be at least

Table 15-2. RAI Staff Schedule (Average Units, Yearly).

	1974	1975	1976	1977
Fully employed staff	11,568	11,530	11,740	11,892
Employees by contract	398	409	373	407

32 hours daily in all of the three radio networks: 16 hours in all on the two line radio channels, which broadcast different programs to those of radio on the air, eight hours in all on the two television networks (six hours in the summer). Advertising must not exceed 5% of the total duration of both radio and television broadcasting.

It is forbidden by law for RAI to broadcast, on radio or television, films of plays for which permission to perform in public has not been obtained, or to broadcast films or plays forbidden to persons under the age of 18 years of age. Various legal provisions specifically regulate the relationship of collaboration between cinema and television Law No. 1213 of November 4, 1975, Article 55), referring back for details to the agreements which are maintained updated between the sectors. There are legal regulations governing both the number of films screened on television every year and the minimum percentage of films and telefilms produced nationally.

Additional services, such as programs for abroad or local programs in minority languages, are explicitly required by the state and are regulated by special conventions.

TECHNICAL UNDERTAKINGS

The current Act of Concession fixes a series of undertakings affecting the technical activities of RAI (Convention of August 11, 1975, Article 110.) On the basis of the overall estimate of the annual revenue of the concessionary and of the finances granted to it by the state, the following outlays (amongst others) of a technical nature have been planned:

● restructuring and extension of existing networks and systems (on December 31, 1976, 1304 television transmitter systems were operating; 128 medium-wave radio transmitters; one long-wave transmitter; 10 shortwave transmitters and 1440 frequency-modulated receivers)

● the establishment of a third television network (a special working group was constituted expressly for this purpose in 1976)

● the utilization of some of the frequencies assigned to Italy by international agreements, a margin being left to cater for private television and radio emitters (provisions of the Constitutional Court Sentence 202, 1976)

● The development of a plurichannel experimental TV network for transmission by cable.

Any projects for new transmitter systems or modification to existing ones require prior approval from the ministry of post and telecommunications, which also oversees any subsequent testing. RAI submits for the approval of the ministry of post and telecommunications both overall technical and financial plans for 3-year periods, and any plans for individual projects. All phases are formulated in terms of the general guidelines indicated by the parliamentary commission. Tables 15-3 and 15-4 list radio and television broadcast installations and population coverage.

CONSTRAINTS ON FINANCIAL MANAGEMENT

A ruling, called the innovating principle, is laid down for financial management. Here, the board of directors and the director general are removed from office when, in any financial year, the total expenditure exceeds by 10% or more the total receipts forecast (Law 103 of April 11, 1975 Article 12). Fairly precise estimates of both the annual receipts and final balances must be reported to the ministry of post and telecommunications as well as to the ministry of the treasury, both of which have the power of control and verification of the estimates (Articles 3 and 4 of the Legislative Decree of the Head of the State Stationary Office No. 428 of April 3, 1974). These powers extend to management as well (Article 7 of Convention of August 11, 1975).

License fees, together with advertising proceeds and any other receipts permitted RAI by law, are required to fit the needs of an efficient and economic management of the radio and television services. To this end, the ministry of post and telecommunications, in agreement with the ministry of the treasury and after consultations with the parliamentary commission, performs a biennial review of the adequacy of the fees. RAI may ask for such a review to be undertaken in advance of the expiry limits (Article 8 of Convention of August 11, 1975).

SOURCES OF INCOME

The main sources of income for RAI are the proceeds of license fees for radio and television sets, radio and television advertising receipts, and reimbursements by the state for special services requested in addition to those covered by the terms of the concession.

Radio Television License Fees

The proceeds from fees constitute the most important source of receipts, both because they are more predictable from an accountacy point of view, and because they are in accord with the logic of the present RAI setup.

Radio and television consumers are required by law to pay a fee for a radio or television license (cf. among others, Royal Decree No. 264 of February 21, 1938) for any sets they may possess capable of receiving, or being adapted to receive, circular radio broadcasting transmitted by air or by cable ("Radio broadcasting" is defined in the international regulation No. 1964 of Altantic City of October 2, 1948, as both the broadcasting of

Table 15-3. Radio Broadcasting Installations and Population Coverage (as of December 31, 1977).

	1974	1975	1976	1977
Medium waves	128	128	128	128
Short waves	10	10	10	10
Long waves				
Frequency modulation	1889	1826	1827	1833
Total number of installations	2028	1965	1966	1972
Population coverage (1)	99.4%	99.4%	99.4%	99.4%

*Calculated on the basis of the population distribution resulting from the 1971 census.

sounds alone (radio listening) and the radio broadcasting of sounds and pictures (or television). Receiving sets for the private consumption of radio and television (both black-and-white and color programs) require licenses, as do family receiving sets, though one radio license and one television license entitles the owner to have more than one of either type of receiver. (Decree No. 557 of December 30, 1946: Ministerial Decree of November 19, 1953; Decree of President of the Republic No. 121 of March 1, 1961; Law No. 1235 of December 15, 1967; Law No. 103 of April 14, 1975; Decree of February 1, 1977, on the initial commencement of relevant telecasts.)

A special license is required for receiving sets (either radio or television or both) put in public places (or places accessible to the public) for public consumption of radio or television programs from them. A family home is not defined as a public place.

Car radio or car TV licenses are required for receivers in cars or motorboats.

The amount payable for license fees is fixed by CIP (Interministerial Prices Committee). Together with the license fee, consumers are required to pay a government concession tax of 1000 lire for radio sets, 4000 lire for black-and-white television sets, and 8000 lire for color television sets (Law No. 1150 of December 10, 1954, and subsequent modifications). The proceeds of the tax accrue to the treasury (Decree No. 633 of October 26, 1972). Value added tax at a rate of 6% is also payable on the receipts of the RAI net of the quotas which are the responsibility of the state administrations. In all, the consumer pays license fees of 3585 lire, 26,170 lire and 52,345 lire, respectively, for the private use of radio sets, black-and-white television and color television sets. Table 15-5 reveals radio and television licenses issued.

These sums are not actually received by RAI. From the amount paid by the consumer, the government concession tax and any quotas retained in favor of various ministries and bodies must first be deducted. Then, from the gross sum thus obtained must be subtracted the monies paid over

Table 15-4. Television Installations and Population Coverage (as of December 31, 1977).

		1974	1975	1976	1977
Network 1	Transmitters	45	45	45	45
	Repeaters	758	770	781	806
	Coverage	*	98.65%	98.66%	*
Network 2	Transmitters	48	48	48	49
	Repeaters	348	396	422	470
	Coverage	*	96.48%	96.56%	*
French and Swiss relay programming in the Valled' Aosta Region	Transmitters Repeaters	— —	1 3	2 6	2 - 10
	Total number number of installations	1199	1263	1304	1382

* Data not available

to the state by RAI in the name of state participation (Convention of August 7 and 8, 1975, Article 28 and sidenote of the state—RAI convention approved the Decree of the President of the Republic No. 452 of August 11, 1975). After the deductions enumerated above, RAI receives approximately 2100 lire for every private radio user's license; 18,618 lire for black-and-white television licenses and 37,281 lire for color television licenses.

For 1977, the net quotas owing the RAI were lower than anticipated, as fee increases only took effect from February 1, 1977. (The figures being 18,248 lire and 35,356 lire, respectively, for black-and-white and color television licenses). Persons caught evading the payment of radio and television license fees and the relevant government concession tax are liable to pay, in addition to the amount of the tax evaded, a fine of two to six times the amount of the fee and the relevant government concession tax. Thus the fine may be as much as 150,000 lire and 300,000 lire, respectively, for failure to pay license fees on black-and-white and color television sets.

Failure to pay either the car radio or television license fee or the concessionary tax, or both, results in a fine amounting to double the license fee payable for a black-and-white car television set, and triple the license fee payable for a color car television set.

Advertising Proceeds

Radio and television advertising, which for RAI constitutes the second source of income after license fees, is regulated expressly by Article 21 of the reform law. Advertising activity is subject, in addition to the limits on percentage incidence on broadcasts and telecasts mentioned above, to further limits deriving from:

● the general orientations prescribed by the parliamentary commission (in accordance with Article 4 of the reform law) designed to protect the consumer and to ensure the compatibility of productive activities with the objectives of general interest and the responsibilities of the public radio and television

● the requirements necessary for the protection of other sectors of information and mass communications.

RAI may undertake advertising directly (Article 21 of reform law), or indirectly by means of SIPRA (Italian Shareholding Advertising Company), whose entire share capital it took over in January 1975. In the interests of smooth functioning, RAI has entrusted to SIPRA exclusively the tasks of acquiring advertisers and formulating contracts. Control and coordination of advertising material for broadcasting and telecasting has been delegated to SACIS (Commercial Shareholding Company for Entertainment Intiatives), whose capital belongs 90% to SIPRA and 10% to ERI. The parliamentary commission, after consultation with the joint commission set up by the presidency of the council (Decree of October 9, 1967, which also covers FIEG (Italian Federation of Newspaper Publishers and RAI)) establishes, on the basis of the levels of national advertising in the press, and on radio and television over the prior year and the current year, the upper limit for advertising revenue. Such a policy is intended to guarantee the balanced development of the two media (Law of April 1975, Article 21, Paragraphs 3 and 4).

Reimbursements for Special Services

Services of social and general utility, which various state administrations call on RAI to provide and which are beyond the scope of RAI's obligations under the convention, are special services and are, as such, reimbursible by the state. As examples of special services for social and political ends there are the radio and television programs in German for the Bolzano Province, the Italian and Slovene programs from Trieste, the French language programs for Aosta Valley, the programs for emigrant workers and the shortwave programs for transmission abroad.

Educational programs are required for schools and occasionally for special groups, such as young national servicemen. Other special services relate to organizational matters (for example, the management of ordinary TV subscriptions), or to technical matters (such as the construction of new plants to serve bilingual border areas in such a way as to facilitate the relay of the programs of foreign organizations).

Table 15-5. Radio and Television Broadcasting Subscription Sales.

	Television			Radio		
	Number	Increase in the year	Desnity per 100 families	Number	Increase in the year	Density per 100 families
1974	11,816,467	389,982	71.3	12,641,302	193,193	76.3
1975	12,102,654	286,187	71.7	12,817,545	176,243	75.9
1976	12,376,612	273,958	72.0	13,024,001	206,456	75.8
1977	12,705,210*	328,598	72.7	13,316,210	292,209	76.1

* of which 399,754 color television

RELATIONS WITH CONTROLLED AND CONNECTED COMPANIES

By law (Article 13, last paragraph of Law of April 14, 1975, and Article 3 of the New Convention) RAI is authorized to undertake collateral activities aimed at promoting the diffusion of its own cultural and artistic productions, or productions in any way connected with its own institutional activities, for both commercial and noncommercial utilization. These activities, done either through its wholly-owned subsidiaries, or through companies in which it has majority shareholding, range from publishing to discography; from audiovisual aids to cinema, theatre and concert utilizations; from the sale of programs and the utilizations of the attached rights to obtaining and utilizing patents through to assistance and collaboration with third parties, providing these do not interfere with the functioning of the company as a public service, and providing the activities do not create imbalance in the running of the companies.

Chapter 16
Broadcasting
In The Netherlands

Dr. Hans H. J. Van Den Heuvel
Government Department of
Culture, Recreation & Social Welfare

The broadcasting system in the Netherlands is unique. Some people will say that its salient feature is its complicated organizational structure; others that its chief distinction is the freedom of expression which it allows. Perhaps both views are correct, since if broadcasting—and therefore programs—are to reflect social and cultural variety as much as possible, this can only be done by means of a fairly complicated system of licensing broadcasting organizations, or at any rate, one more complicated than if there is simply one national organization. When as many groups from society as possible are given the opportunity to participate in radio and television broadcasting and, moreover, to establish their program policy completely independently, then it can be said that the principle of freedom of expression in broadcasting is being put into effect to a very great extent.

THE DUTCH BROADCASTING SYSTEM:
COMPETING FOR THE SUPPORT OF THE PUBLIC

However one chooses to describe Dutch broadcasting, one thing is certain: it is an open, pluralist system. By law, organizations may become licensed broadcasting organizations if they enjoy a certain degree of support from the public; depending on the actual level of that support, they are allocated broadcasting time on radio and television. The majority of Dutch broadcasting organizations are, therefore, private companies. The Dutch broadcasting system has its roots in the cultural and social history of Dutch society. Its structure has always been dominated by the question whether there ought to be one all-embracing organization (within which the diversity of Dutch life would have to find expression) or whether that diversity itself should constitute the basic principle, implying a multiplicity of organizations corresponding to the most important groupings in Dutch society. From the earliest days of broadcasting (the mid '20s) to the present, the second of the two solutions has prevailed, though the debate still continues.

THE BEGINNINGS OF RADIO BROADCASTING

On November 5, 1919, a national newspaper, the *Nieuwe Rotter-damsche Courant* contained an advertisement announcing the first radio broadcast ever to be made in the Netherlands. The broadcast, entitled *Soiree Musicale*, was to take place the following evening and anyone possessing a "simple radio receiver" would be able to tune in. It was made by the Netherlands Radio Industry Company, a small factory producing radio-transmission and receiver equipment. Regular broadcasts were made with the factory's transmitter from 1919-until the end of 1924.

The owner and director of the factory (Fig. 16-1), Hanso Henricus Schotanus a Steringa Idzerda, was born in Weidum, a small village in the province of Friedland in September, 1885. His parents had expected him to maintain the family tradition and become a clergyman or a doctor, but this was not to be. Idzerda showed a special interest in technology, and in radio technology in particular. In 1913, he settled in Scheveningen, near The Hague, as an advisor on all applications of electricity.

What had once been his hobby, now became his profession, and he began to occupy himself intensively with wireless telegraphy, the development of equipment for the transmission of signals without wires.

As a result of his experiments, he succeeded in developing a radio valve (vacuum tube) which was produced according to his instructions by Philips of Eindhoven and which appeared on the market toward the end of the First World War. Gradually, the thermionic valve was largely to replace the crystal receiver, and together with the transmitter, which he built himself, it enabled Idzerda to make radio-telephone transmissions; this was something new, as previously only wireless telegraphy (usually Morse code) had been possible.

During a trade exhibition in Utrecht early in 1919, Idzerda rigged up an experimental radiotelephone connection over a distance of approximately 1200 meters. Some people suspected the presence of an underground cable, but reports from amateur radio enthusiasts all over the country who had picked up the conversation on simply homemade equipment soon refuted this allegation. This historic event, which was witnessed by Queen Wilhelmina herself, caused considerably excitement among radio "hams."

After his first official radio program on 6 November, 1919, Idzerda made regular broadcasts several evenings a week. But radio broadcasting was an expensive business, and after a few years he got into financial difficulties. In 1922, when his money problems were threatening to become serious, an appeal in an English magazine for radio enthusiasts, *Wireless World*, raised £750. The same year, the popular English newspaper, *The Daily Mail*, decided to finance the weekly concerts which Idzerda broadcast from the Kurhaus in Scheveningen and which became well-known in England as the "Dutch Concerts." Nevertheless, he was obliged to make repeated appeals to his listeners for financial support as program and broadcasting costs continued to put heavy pressure on the operating results of his business. When the British Broadcasting Company, Ltd., was founded to provide the British with their own radio

Fig. 16-1. Ir. Hanso Henricus Schotanus A Steringa Idzerda (1885-1944), one of the early pioneers of Dutch broadcasting.

broadcasts, financial support from England ceased. Idzerda's broadcasts came to an end (November, 1924), and the Netherlands Radio Industry Company went bankrupt.

THE RISE OF THE BROADCASTING ORGANIZATIONS

Before Idzerda's transmitter fell silent, another enterprise, which also made an important contribution to the development of broadcasting in the Netherlands, had started broadcasting radio programs, the Neder-laddsche Seintoestellen Fabriek (NSF), the Netherlands Transmission Apparatus Factory. The NSF was established early in 1918 by a number of Dutch shipping companies to produce navigation equipment such as ships' compasses and radiotelegraphic transmitters and receivers.

After the First World War, this factory, whose customers had previously been largely from overseas, was forced by the economic situation to concentrate on radio equipment for the home market; therefore, it became a serious competitor to Idzerda's Netherlands Radio Industry Company. The Netherlands Transmission Apparatus factory also decided to have its own transmitter built (by the English engineer G.W. White) and to make broadcasts to boost sales of its radio equipment (Fig. 16-2). The NSF began its broadcasts in mid-1923 and was soon to leave Idzerda far behind as regards both the technical aspects of sound quality and the budget available to it for making programs (Fig. 16-3).

Nevertheless, as time went on, the experimental broadcasts cost the NSF a great deal of money, even though it was financially much stronger than Idzerda's company, primarily because of the increasingly professional organization of the programs and the fees paid to the growing number of staff and artists. Therefore, in March, 1924, on the initiative of the director of the NSF, Antoine Dubois (who was to become very well-known prior to the Second World War, notably as president of the Union Internationale de Radiodiffusion), a separate organization was set up, the Hilversumsche Draadlooze Omroep (HDO), the Hilversum Broadcasting Organization, which took over programs and their financing. An appeal was made to listeners to make periodic contributions, which entitled them to membership of the organization. Renamed the Algemeene Vereeniging Radio Omroep (AVRO), the General Radio Broadcasting Organization, sometime later, it was thus the first broadcasting organization in the Netherlands.

The AVRO's broadcasts were intended for a general audience. The programs provided within the limited broadcasting time were, therefore, varied, consisting primarily of entertainment, with an occasional lecture. The AVRO thought it could take up an independent position by adopting what was seen as a "neutral" program policy: the programs would include something for everybody. In this way, it hoped to reach as many different groups of the population as possible, without becoming identified with one or another social or religious school of thought.

When a second transmitter was built by the NSF in 1927, the new broadcasting time was divided between the two large denominational broadcasting associations, apart from a small part which was allocated to the VPRO Broadcasting Association, a small organization with liberal protestant leanings, set up on 1926. By this time, the pluralist structure of Dutch broadcasting was a fact, and it was then the task of the Government to provide it with a legal basis.

GOVERNMENT POLICY

The government had not, up to this point, concerned itself very much with broadcasting, limiting its involvement initially to issuing factories

Fig. 16-2. Ir. G.W. White, constructor of the first transmitter in the Netherlands (circa 1920).

with licenses to broadcast, to allow them to experiment with new inventions and to test equipment. However, with the development of private organizations, which had no connection with the factory licenses and which were devoted entirely to the production of radio programs, the government decided to take charge of the licensing of the organizations, rather than leave the question of who was allowed to broadcast to the commercial practices of the factories, which made their transmitters available to third parties in return for a fee. The government, therefore, laid down conditions to be fulfilled by organizations which wanted to broadcast radio programs.

In 1928, the first legislation supplementing the 1904 Telegraphs and Telephones Act was enacted, providing for measures to be taken, where necessary, without actually establishing any rules concerning broadcasting. Two years later, in 1930, the government introduced a number of important legal provisions which can be divided into three categories.

First, conditions with which broadcasting organizations had to comply if they wanted to be eligible for a broadcasting license. They had to have the legal status of a corporation, to aim at the satisfaction of cultural or religious needs existing amongst the population and thus to be in the public interest.

Secondly, the government divided the available broadcasting time on the two radio stations equally between the existing broadcasting organizations so that each of the four large religious/political associations (AVRO, KRO, NCRV and VARA) was allocated approximately one quarter of the time. Each of the broadcasting associations produced programs including plays, music, light entertainment, children's and women's programs, as well as informative and education programs which clearly reflected the religious and ideological atmosphere prevailing amongst the supporters for whom the programs were intended. Because the associations directed

291

their programs almost exclusively toward their own groups of the population (who were at the same time their members) and because the link between an association and its adherents was very strong, the character of the system was very much that of a "closed shop" and its organization was very inflexible. The existing associations jealously resisted any encroachment on their autonomy.

But in one important respect—the third government provision—the broadcasting organizations did have to tolerate outside intervention: the government was to be responsible for ensuring that programs did not contain anything which could threaten national security or undermine public order or morals. A special committee was set up to carry out censorship on behalf of the government.

The pluralism of the Dutch broadcasting system was limited largely to the four large broadcasting organizations, each representing a particular school of thought in society: Protestants, Catholics, Socialists and one group which was not politically or religiously committed.

After the Second World War, the closed-shop character of the broadcasting system was retained, although the organizations cooperated in the Netherlands Radio Union (NRU), on whose premises they had their studios, technical equipment and music libraries and which was responsible for providing a joint program. Censorship was abolished after the war, but the government became much more involved in the running of broadcasting when it appointed a government official to supervise the financial side. The most important reason for this was that, whereas before the Second World War the broadcasting associations had been financed by voluntary contributions of their members, after 1945 they received funds levied by the government as a wireless license fee.

TELEVISION

After several years of experimental broadcasts, the government authorized the existing broadcasting associations to broadcast television programs starting in October, 1951 (Fig. 16-4). Here, too, the associations founded a coordinating body, the Netherlands Television Association (NTS) and the government collected revenue in the form of a television license fee in order to finance programs. The organization of the NTS was analogous to that of the NRU for radio, and its aim was to create the most efficient cooperation possible between the associations with respect to the allocation of broadcasting time, the coordination of programs and the general operation of the studios. The NTS was also responsible for the joint programs.

The allocation of radio broadcasting time by the government in 1930, after years of controversy, was highly significant since it meant that each of the various groups of the population was to be granted an equal amount of broadcasting time and that the Netherlands would have a pluralist system made up of a number of autonomous organizations rather than one overall national broadcasting organization. The issue of whether commercials should be introduced on television provoked a similar controversy during a subsequent period of Dutch broadcasting history.

Fig. 16-3. Transmitter of NSF with constructor, G.W. White.

ADVERTISING

The question of advertising arose when the government was considering plans to introduce a second television channel in order to increase the choice for the public and the broadcasting potential for the broadcasting organizations. At the same time, various commercial initiatives developed with the aim of penetrating the closed-shop broadcasting system and exploiting television for commercial ends. The business world had been aware, for some time, of the advertising potential of television to boost sales of products. Making the commercials would also be highly lucrative. The success of British businessmen in breaking the monopoly of the BBC, which led to the introduction of commercial television in Britain, inspired the idea of a similar arrangement in the Netherlands. The business world, including industry, the banks, insurance companies and daily newspapers, wanted a system whereby the broadcasting associations would continue to provide the programs on the first channel while it would operate the second channel on a commercial basis.

The battle for the right to television advertising was played out primarily between a number of fiercely competitive groups of financiers and made little impression on the general public.

Parliament, however, was not indifferent to the question of whether commercial television should be introduced in the Netherlands. It twice rejected government proposals on this point; as in the 1920s, the majority of members of parliament wanted broadcasting to be left in the hands of organizations representing groups of the population and not to be exploited for commercial ends. Therefore, it was decided that the second television channel would also be run by the broadcasting organizations, rather than commercial enterprises. However, it was decided to allow radio and television advertisements on a limited scale before and after news broadcasts (as from 1 January, 1967 on television and 1 March, 1968 on radio).

Advertising became the exclusive responsibility of the Television and Radio Advertising Organization (STER), a government body, and all profits are ploughed back into broadcasting (from 1967 to 1973 a percentage was paid out to newspapers and magazines as compensation for lost advertising revenue).

At the same time, the government decided that other groups of the population should also be given the opportunity to broadcast radio and television programs, in addition to those represented by the existing broadcasting associations; in other words, the closed-shop system should be scrapped without deviating from the fundamental principle of freedom of speech. After a short transitional period from 1965, the 1967 Broadcasting Act became effective in 1969 and provided the Netherlands with a new broadcasting system.

THE BROADCASTING ACT

The most important innovation contained in the Broadcasting Act was that new organizations could be granted a license, provided that they had a certain minimum number of subscribers to the program guides (to which each broadcasting organization holds the copyright) or people who are not regular subscribers to a program guide but who make an annual financial contribution to one particular organization. The available broadcasting time is divided between the organizations, including the original ones, in proportion to their numbers of members.

The Broadcasting Act establishes categories of broadcasting organizations for the purpose of allocating air time: Category A, at least 400,000 members; Category B, between 250,000 and 400,000 members; and Category C, between 100,000 and 250,000 members. The Act also provides for aspirant broadcasting organizations, which must have at least 40,000 members and must achieve Category C status (at least 100,000 members) within two years. The broadcasting time available for categories A, B and C is divided up in the ratio 5:3:1. Aspirant broadcasting organizations receive a weekly allocation of three hours of radio time or one hour of television time, or in some cases, both, for a maximum of two years.

The authorized broadcasting organizations must provide a varied selection of programs, in principle, of all categories including informative, educational and cultural programs and entertainment. The Broadcasting Act contains a condition which has, in fact, applied since 1930 and which determines the basic character of the Dutch broadcasting system: the activities of the broadcasting organizations must be aimed at satisfying cultural, religious or spiritual needs of the population so that the broadcasts are of benefit to the general public. The organizations must not be oriented toward or instrumental in making a profit.

The Broadcasting Act also created a new organization, both to coordinate all the other broadcasting organizations and to broadcast its

Fig. 16-4. The first TV studio in the Netherlands in the 1950s, established in a former church.

own programs. The Netherlands Broadcasting Corporation (Nederlandse Omroep Stichting, NOS), as this body is called, was founded in May 1969 to replace both the Netherlands Radio Union (NRU) and the Netherlands Television Foundation (NTS). Its responsibilities include providing program facilities (studios, sets, graphic designers, etc.) for all the organizations to which broadcasting time is allocated, coordinating programs and supervising the conditions of employment of all broadcasting personnel. It also provides certain types of programs which a coordinating organization can best provide, such as news and sports programs and programs expressing the views of sectors of the population which are not catered to by the other broadcasting organizations. Finally, the NOS represents Dutch broadcasting abroad, notably in international broadcasting organizations.

Some time ago, parliament decided to introduce a new condition for organizations that wish to become aspirant broadcasting organizations. In the future they will be accepted only if they have something essentially new to add to the existing choice of programs. It is yet unclear how the new criterion "something new" is to apply in a system already based on diversity. The decision has not yet come into force (as of June 1979), nor has a government proposal, approved by parliament, to increase the minimum number of members for the different categories of broadcasting organizations. The new figures are as follows: Category A, more than 450,000 members; Category B, at least 300,000 members; Category C, at least 150,000 members; and aspirant broadcasting organizations, at least 60,000 members.

Three new organizations have now become established in Dutch broadcasting, in addition to those already mentioned: The Evangelische Omroep (EO), the Evangelical Broadcasting Association which has a very orthodox denominational character; the TROS Broadcasting Association, which has no denominational or socio-political leanings and is neutral; and the Veronica Broadcasting Association (VOO), whose programs are directed mainly at a younger audience.

The AVRO, KRO, NCRV and VARA have suffered as a result of the competition provided by the new organizations. In order to retain the support of the public, they have devoted less attention to programs embodying their own principles and concentrated more on entertainment which appeals to a mass audience. Furthermore, the link between the broadcasting associations and the groups of the population from which they originated is no longer so self-evident. The public identifies with the broadcasting organizations less than in the past, in keeping with general developments in society such as crises of identity, superficiality, the absence of accepted standards and insecurity.

Finally, the open-door approach has weakened the position of the established broadcasting organizations through the creation of the NOS. Whereas the NRU and the NTS were, to a large extent, controlled by the existing organization, the NOS is an independent broadcasting organization with its own staff and advisory councils and broadcasting time on radio and television. Since, it is also the coordinating organization , the NOS is not required to have members. Its allocation of broadcasting time is laid down by law; it does not have to worry so much about competition from the other organizations, as is often evident from its programs which are based less on the tastes of a mass audience.

The antenna facilities of a modern Dutch TV station appear in Fig. 16-5.

THE DUTCH WORLD BROADCASTING SERVICE

Dutch broadcasting not only involves competing for the attention of the public at home. Efforts have been made since the early history of broadcasting to attract listeners abroad. In 1927, regular shortwave radio

Fig. 16-5. Radio/TV transmitter in the Dutch landscape.

programs were transmitted from the Netherlands to very distant countries, notably by Phillips in Eindhoven to the former colonies in the Dutch
East Indies (now Indonesia). Phillips established a special broadcasting
organization for this purpose, the Philips Holland-East Indies Broadcasting Organization. Ten years later, programs were also being broadcast to
Suriname, the Netherlands Antilles, North America and South Africa.

It was after the Second World War that world broadcasting activities
in the Netherlands really came into their own. To begin with, in 1947,
responsibility for overseas broadcasts was given to a specially created
government organization, since it was felt that the financing of the

programs would not then be dependent of the government and has an independent board and program advisory council whose members come from a wide variety of backgrounds.

The programs themselves also developed considerably after the Second World War. Before 1939, all broadcasts were made in Dutch, being intended, exclusively, for the Dutch colonies, Dutch people abroad and the Afrikaans speaking people of South Africa. Nowadays, the Netherlands World Broadcasting Service provides direct daily broadcasts in Dutch, English, French, Spanish, Portuguese, Indonesian and Arabic.

The pre-war broadcasts were intended primarily to provide Dutch people abroad with a link with their native country. Those of today aim to present a picture of the Netherlands to the rest of the world, showing its spiritual, moral, constitutional, cultural, scientific, economic, social and humanitarian characteristics. The World Service also has a responsibility to promote goodwill toward the Netherlands and to contribute to the establishment and maintenance of peaceful international relations and international cooperation.

Besides its direct daily broadcasts, the Dutch World Broadcasting Service also sends programs both for television and radio on record and tape to other broadcasting stations. These are provided free of charge, mostly to stations in Third World countries and to relatively small, noncommercial stations in America.

Finally, the Dutch World Broadcasting Service has its own international training center, where annual courses in radio and television broadcasting are provided for about 50 people working for broadcasting organizations in Africa, Asia and Latin America. The courses are not always held in the Netherlands, but sometimes take place in the country or region of the participants themselves. They provide training primarily in compiling and producing information and education programs.

Chapter 17
Broadcasting in Sweden

Herbert Soderstrom
Information Director
Swedish Broadcasting Corporation

The average Swede spends almost two hours each day in front of his television set. And that, we know from surveys on family patterns, is more time than most parents spend talking to their kids. Swedish audience research is carried out by a department within the Swedish Broadcasting Company, but which has an excellent international reputation—especially when it comes to viewing habits among children. The method is mainly field interviews sampled in the same way Harris or Gallup do.

At the age of nine the young Swede is an established TV viewer; the kids spend on average 111 minutes a day in front of the screen. This goes on till the age of 14 when, with the discovery of the other sex, their TV habits change drastically. In the 15 to 24 age group TV viewing habits drop to less than half; 55 minutes a day. Then slowly the Swede resumes his viewing habits so by the age of 65 he (or she) is back to where he started, at about 110 minutes a day workday. On the weekend, people tend to watch somewhat more.

Radio habits are, roughly speaking, in inverse proportion. Those who watch TV less than average tend to listen to the radio somewhat more. Young parents, for instance, spend relatively little of their time in front of the TV, only 76 minutes, but listen to the radio for three hours on workdays. Average radio consumption is the same as the time spent on TV, about two hours per day. (See Fig. 17-1).

BROADCASTING ORGANIZATION

The output of radio and TV is spread over three radio channels and two TV channels. The daily output of radio is 50 hours, and TV 12 hours, mainly in the evenings. All channels are strictly non commercial. The total production cost, 1 million SwKr (220 million $US) is covered by a license fee for each household owning a TV set; radio is free of charge. A color TV license costs $100 per year.

299

Program production and buying is, since July 1, 1979, shared by four production companies, fully owned by an "umbrella" company, the Swedish Broadcasting Company. The companies are: Sveriges Television AB, responsible for the three radio channels; Sveriges Utbildingsradio AB, which produces educational programs for both radio and television; and Sveriges Lokal-radio AB, which administers 24 regional stations. (Figure 17-2 is a typical scene in a busy radio studio.)

The Swedish Broadcasting Corporation (Sveriges Radio AB) is the economic center of the broadcasting industry in Sweden, while the subsidiaries are fully responsible for the programming. This double command has been vividly discussed, but as long as the new organization is still in its first fiscal year, no ultimate answer can be given to the questions raised about the concentration of power within the media.

The "umbrella" company itself, the SBC-SR, is a regular private company, although noncommercial. It is owned by three different shareholding groups, the press in Sweden (20%), various entrepreneurial organizations (20%), and a wide spectrum of voluntary organizations, ranging from trade unions to churches and teetotallers (60%).

The license fee paid by each TV household is collected by the Swedish Telecommunications Board and stored in a license fund. One year in advance, the Swedish Broadcasting Corporation puts in its demands for the coming fiscal year. The Government of Sweden presents a bill to parliament, which allocates a lump sum for the coming year. How the money is distributed is left entirely up the Swedish Broadcasting Corporation. This is to ensure that the subsidaries remain free from any sort of political pressure.

In recent years, the Government has felt rather free to express its views about programming, economic efficiency and expansion, so even if the SBC-SR is formally free to make any economic decision it wants within the limits of the lump sum, there is no doubt about what the government's views are. The situation is probably best described as a precarious balance of power.

The board of governors in SBC-SR is partly elected by the shareholders and partly appointed by the government. The two groups match each other in numbers, and two representatives from the employees complete the board.

During the last decade very few votes have been taken by the board, and on those few occasions when they have, the votes have never been cast along group lines. One reason for this is that the government appointees are selected from the different political parties, including those in opposition. There is very little evidence of governmental influence through the board of governors.

PROGRAM CONTENT CONTROL

Program content is controlled by the Radio Council (Radionamnden), a standing state committee, which supervises both radio and TV. The Radio Council is open for complaints from anyone, even foreigners are entitled to file complaints about programs and some embassies have done

300

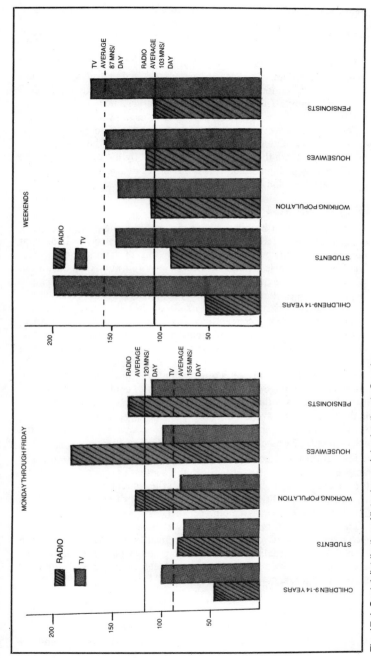

Fig. 17-1. Social distribution of listening and viewing time in Sweden.

so. The council works similar to a court, examining the programs to see if they are in accordance with the Swedish Radio Act. However, the council can only reach a decision over programs already aired—censorship is explicitly forbidden.

The radio act contains two important keystones: *Factuality* and *impartiality*. Freedom of speech is guaranteed, but the programs must promote the ideas of democracy, the equality of humanity, and individual freedom and dignity. The radio council has no judicial power other than to determine whether or not a particular program concurs with the radio act.

The council delivers an annual report to the government, in which it advises the government whether or not to prolong the transmitting license. If the transmitting license is withdrawn by the government, the corporation must be liquidated. This is obviously such a drastic step that it can hardly be taken, and it has never ever been suggested.

POLITICS AND FAIRNESS

The Vietnam War brought about a major test of the program content guidelines with regard to the radio act. SBC-SR was accused from the right of favoring pro-FNL news and from the left of favoring the South Vietnamese and American points of view. The corporation then decided to have all of its Vietnam material investigated by the political science department at the Gothenburg University over a period of several months.

The research group, headed by Professor Jorgen Westerstahl, soon found out that the two key conceptions, factuality and impartiality, were of little relevance when it came to judging program material. Instead they tried additional definitions as follows: factuality contains an element of *truth*, but obviously one can stockpile true statements without being factual, e.g. by consistently selecting true but one-sided facts. An element of *relevance* is also needed. Similarly, impartiality contains an element of *balance*, which means that all parties involved in a conflict are entitled to give their views or have their views presented. And when presenting these views the program must be neutral: so *neutral presentation* is the fourth element. The model will look as shown in Fig. 17-3.

This truth-relevance model was used for the research, and the findings gave evidence that SBC-SR in the Vietnam conflict had in actual fact remained faithful to the radio act. The model has been widely discussed amongst political scientists, and several critics have pointed out various weaknesses. But as a method of analysis it is still in use.

The combination of the fairly liberal radio act stressing freedom of speech, the ideas of democracy, the equality of humans, and the interpretation of factuality and impartiality tends to favor a tolerant and educated outlook, and this can also be seen in the production and consumption figures.

AUDIENCE RESEARCH

Over the years the Audence Research department has collected data on viewing and listening habits with regard to program content, and the

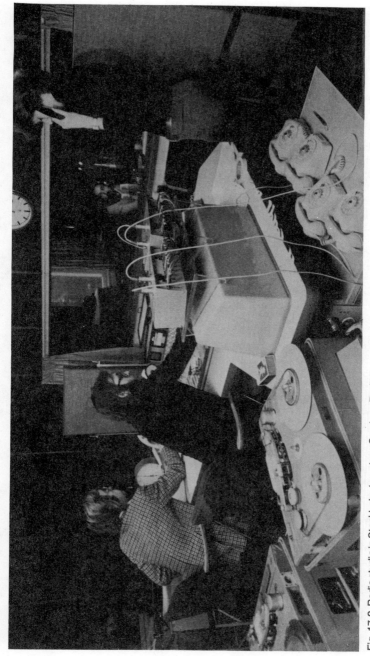

Fig. 17-2. Radio studio in Stockholm (courtesy Sveriges Radio).

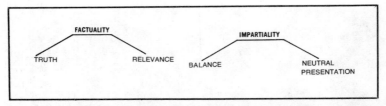

Fig. 17-3. Neutral presentation elements, as seen by the Political Science Department, Gothenburg University.

output figures have been analyzed accordingly. Table 17-1 conveys audience preferences for the main television program categories with the share of program time devoted to each category. The categories favored by the audience are obviously news, theatre, entertainment and sport. The categories favored by SBC-SR are current affairs, culture and religion, and children's programs. For radio the figures are even more revealing (see Table 17-2).

These discrepancies have been recorded over several years and there is no sign of any willingness within the SBC-SR to adapt to audience preferences. This is possibly the most important difference between a commercial and a noncommercial system: the noncommercial system tends to promote a slightly more intellectual ideology in "good taste": current affairs, culture and religion, lectures and serious music, whereas what the public wants is entertainment, light music and sport.

This broadcasting attitude is sometimes described as a paternalistic way of treating the audience, as against the commercial approach which tends to adapt itself to consumer habits. The cynics within the Swedish Broadcasting Corporation still chuckle over an intramural incident that occured when the late King of Sweden died. The two TV channels convened to see if any of the programs scheduled from the time of the King's death till his funeral a week or so later, were too lighthearted. Channel 1 at first sight could find no necessary schedule change, while Channel 2 after some consideration decided that the Liza Minelli show ought to be postponed (when you saw it six months later, you could wonder why; it wasn't that lighthearted). And the result, say the cynics, is that Swedish television is usually so deadly serious that it *always* concords with national mourning.

The other side of the coin, of course, is that television and radio in Sweden do have a reputation with the general public for seriousness, reliability and for having a strong integrity. Television is never looked upon as the "telly," as cheap entertainment for the masses. TV in Sweden is a political factor of great importance, and its cultural impact is immense. (Fig. 17-4 is a view of a TV studio in Stockholm.)

The Kingdom of Sweden with its eight million inhabitants is a national state with very few traits and federalism, and this national unity is reflected in its radio and television. Of the total TV output 97% is national; only 2.5 hours per week are regional transmissions. There are no local TV stations in Sweden. Eighty-four percent of the radio output on all three

304

channels is national; the rest is mainly regional, although there are some experiments with local radio. During the spring of 1979, some small local radio stations came into being, run by local churches, trade unions and other organizations on a strict noncommercial basis. These stations are not required to be impartial. On the contrary, the idea is that they shall be partisan stations, giving each organization the opportunity to air its message.

There is, of course, programming capacity outside of Stockholm, the Capital of Sweden. But the outlet remains, however, national to a very great extent. A third of all TV and radio programs are regionally produced but broadcast nationally.

IMPORTED PROGRAMS

Small nations tend to rely, to a large extent, on imported TV programs, and Sweden is no exception: almost half of the first-run output is imported or bought from independent production companies in Sweden. The SBC-SR imported 1056.8 hours from abroad (sports excluded) in 1978, of which the main exporting countries were:

United Kingdom	297.0 hours
USA	250.6 hours
France	105.0 hours
West Germany	76.5 hours
Canada	41.8 hours
Other countries	285.9 hours

Among "other countries" the Nordic neighbors (Denmark, Norway, Finland and Iceland) contribute the most. Imports from Socialist countries are almost insignificant.

As one can see, the imports pour in almost exclusively from the Western Hemisphere, and many of the imported programs are originally produced by commercial companies. The Swedish audience is as familiar with "Rhoda," "Nancy" and "Holocaust" as any American TV viewer. Fiction programs are expensive, especially serious drama, and the result has been that the Swedish TV audience today is more exposed to foreign drama than to domestic.

**Table 17-1. Comparison of Television Program
Content and Audience Viewing Preferences.**

PERCENT OF TOTAL PROGRAM OUTPUT	PROGRAM CATEGORY	PERCENT OF TIME SPENT VIEWING
12	News	19
9	Current Affairs	4
8	Culture and religion	3
3	Nature and science	3
6	"Magazines"	7
19	Theatre and entertainment	34
8	Cinema film	9
4	Music	3
12	Sport	16
13	Children's programs	2
6	Miscellaneous	1

What effect this will have on the national identity is very much an open question. It is possible that Sir Lawrence Olivier can portray Hamlet better than any Swedish actor. But nevertheless, a nation still needs to experience the treasure of world literature in its own mother tongue, otherwise the national identity threatens to get lost.

In addition to that, all imported TV programs are subtitled, not dubbed. This is a normal procedure in a small country, as dubbing is an expensive and elaborate procedure. Moreover, subtitled films were already an established fact in the cinemas when TV started in Sweden, so that SBC-SR merely inherited an accepted practice. This leads to a challenging linguistic situation: more than a quarter of the TV output in Sweden is in English! No wonder that the youngsters are so "tuned in to" English that the teachers of the Swedish language and history are worried that in a generation or two the mother tongue might vanish!

But subtitling is also used as an aid for the deaf, and some of the Swedish TV programs are subtitled—in Swedish. The new technical device making it possible to transmit TV text separately and simultaneously is now being studied to find new ways of using subtitles for the benefit of linguistically handicapped groups—the deaf, immigrants and so on.

The paternalistic radio and TV system in Sweden reflects many examples of a noncommercial, almost anticommercial attitude. Audience maximization is not the goal, education and service also to relatively small minority groups is essential. Twenty-five percent of all TV programs are seen by audiences of 4% or less—and this in a two-channel system where a particular program can be seen by 70% or more of the audience (very popular Swedish cinema films, Swedish-produced theatre, tennis when Bjorn Borg is winning the American Open, etc).

A good example of this minority-oriented programming is the immigrant service. One tenth of the Swedish population today is of foreign extraction in the first and second generation; thus, radio and TV give a fair share of the air time to them: more than four of the weekly 80 hours of TV is for immigrants, mainly in their respective language. More than 10 hours of the total national weekly radio output is in Finnish, Greek, Turkish, Serbo-Croatian and other immigrant languages, while Laps, the "aboriginees" of Northern Sweden, have their own regional radio transmissions.

PROGRAM VIOLENCE AND SEX

For a long time, the Sweden Broadcasting Corporation has been very sensitive to violence. Violence in news and current affairs is accepted as part of the reality TV is there to describe, but fictional violence is only permitted if it has a meaningful, artistic or intellectual function in the story. But even then, the violence is kept under strict control.

Children's programs are extremely restrictive when dealing with violence. All the popular American cartoons that treat violence as something funny are out of the question. The main rule is: violence as entertainment is unacceptable.

Table 17-2. Comparison of Radio Program Content and Audience Listening Preferences.

PERCENT OF TOTAL PROGRAM OUTPUT	PROGRAM CATEGORY	PERCENT OF TIME SPENT LISTENING
12	News	14
13	Lectures, Features	5
3	Religion	2
8	"Magazines"	18
2	Theatre	1
6	Entertainment	11
31	Light Music	44
21	Serious Music	2
2	Sport	2
2	Miscellaneous	1

Sex and nudity are dealt with in a looser way, but also with certain restrictions. In Sweden, as with all other Western countries, sex and nudity do upset some people. The information department of SBC-SR noted a few years ago how many complaining telephone calls each breast of a naked woman gave rise to: 26 calls per breast!

"COMPETITIVE" PROGRAMMING

The paternalism of SBC-SR has influenced programming in a way that is almost unknown in other countries, with the result that peak TV time in Sweden is limited to only one hour each evening. Until 1969, there was only one TV channel. With the introduction of TV 2, the number of TV hours doubled, and roughly speaking each program category was also doubled, since the two channels are each obligated to provide a whole range of programs. As a result, the news coverage was automatically doubled. But, the audience analysts soon found out that with the start of TV 2, the news consumption was reduced to half; in other words, when the output doubled, the consumption was halved.

The explanation of this phenomenon was that people, when provided with an alternative, tended to choose something other than the news. This pattern was strengthened by the program strategy that the two channels adopted to begin with, the so-called "contrast principle": facts contra fiction, broad majority programs contra minority service programs, etc. So when one channel carried the news, the other channel had some broad fiction. And since people tend to prefer broad fiction, the news program lost heavily. A side effect of this symetrically planned channel system was that both channels lost their identity, and people afterwards could never remember on which channel they had seen what.

So the two channels decided unanimously to "protect" the news. Instead of broad fiction contra news, minority service now became the alternative. The manipulation was successful, and now the news audience figures are back to about 20% for each channel. But this "protection" inevitably led to a loss in peak viewing time, which was now reduced to the time span between the end of the news on Channel 2 and the beginning of the news on Channel 1, which happens to be the hour from 8 to 9 p.m. and, alas, gone was the rest of the peak viewing time. This gives rise to a perpetual general complaint over "program collisions" between the two channels. But in a "paternalistic" system there is no other way of solving the problem: how to convince, seduce and force people to take in reality when fiction is so much more tempting.

Within the new organization of the company, radio and television are partly different, partly similar. Different is the channel conception; the three radio channels differ with regard to content and character.

Channel 1 is mainly devoted to news, lectures, current affairs, radio theatre, and various types of music. In a way, it resembles the BBC Home Service. This is the channel for the educated. It runs from the early morning till about 11:00 p.m. in the evening.

Channel 2 was originally conceived for serious music, but has gradually developed into a combined music and minority service channel, and most of the educational programs are now located on this channel. It starts about 6:00 in the morning and generally closes down at midnight. Its orientation toward minority groups as well as its serious character make it the least listened to, and its audience figure seldom rises above 2%.

Channel 3 was started as a counter move to a commercial "pirate" station which was broadcast from a ship in the Baltic Sea and consequently carried mainly news headlines and light music. As a result of joint Nordic legislation, when the "pirate" was forced out of business, Channel 3 slowly changed: light entertainment, traffic programs, more extensive news coverage, and, in later years, the expanding regional radio now making up the bulk of its contents.

All three channels are scheduled for simultaneous transmission. One principle is that the listener shall always have at least one musical alternative which is generally "easy" to listen to, light music, folk music, operettas, and the like. Strictly speaking, there is a difference between *channel* and *program*.

The channels are defined in a technical sense, and thus 99.8% of the country is served by FM stations which transmit all three channels. The transmitters are operationally run by the Swedish Telecommunications Board, while the program scheduling is the responsibility of the radio company, the Sveriges Riksradio AB.

But program content responsibility lies with the various program production companies. The radio company is in itself the biggest program contributor, with about 60% of the first runs. Regionally produced programs for national distribution occupy about 30% of the radio channels, the Educational Radio Company (Sveriges Utbildningsradio AB) and the Regional Radio Company (Sveriges Lokalradio AB) are responsible for about 5% each.

Legal responsibility lies exclusively with each production company. A separate Radio Responsibility Act regulating criminal offenses in radio and television (libel, slander and the like) requires one single prenominated editor for every program heading, and the production companies are responsible for ensuring that a log is kept to cover any possible incident with regard to the Radio Responsibility Act. In addition to this, the government and each production company have written agreements covering the scope and intentions of the respective companies.

Fig. 17-4. TV studio in Stockholm.

This rather complicated network of regulations and responsibilities is intended to ensure decentralization, diversity and integrity in either media in Sweden. The best example of how it works is the agreement made by the National Radio Company and the Regional Radio Company. The scheduling power lies, as mentioned above, with the National Radio Company. It allocated three time slots for the Regional Radio Company on Channel 3 in the morning, at lunch time and in the early evening. Channel 3 is then split up into 24 (sometimes 26) regions, each with its own station, fully responsible for what is put on the air. But once its time is up, the regional station must end its transmission and hand over to the national network.

If a regional station urgently needs some more time on the air, it can ask for it from the National Radio Company, a procedure which is not unusual. The Regional Radio Company has its seat in Stockholm, but the 24 regional stations are to a very large extent autonomous in their program planning; the economy on the other hand, is entirely in the hands of the central company. This is the general feature of the whole structure of Swedish Broadcasting Corporation, i.e., centralized economic planning, scheduling on a more decentralized level (parallel to the "umbrella" company) and program decisions made very close to the everyday production.

The TV organization is different. The two channels are of equal importance, both with about 40 hours of weekly output. Both are required to provide full coverage, meaning that both have to carry all program categories. They are kept together within the Swedish Television Company, Sveriges Television AB. The economic power (within the framework set down by the "umbrella" company) lies with the television company. But, unlike the radio, each channel is responsible for its own scheduling. A tight network of voluntary regulations (such as "protection" of news, already mentioned) makes it hard to say that the scheduling is all that independent. The two channels differ very little in program strategy, and if there are any differences worth mentioning, one can point out that Channel 1 puts on 35 hours more per year of nature and biology, whereas Channel 2 takes the lead in foreign affairs, with 40 hours more per year. The relation between the different program categories is very stable, and one can find exactly the same percentages year after year. Table 17-3 is a profile of Swedish TV programming.

The internal organization within the two channels of TV is somewhat different. At the head of each channel is a so-called program director, who, in TV 1 has an administrative staff and a program secretariat. TV 1 has come to terms with the ideas of the new organization and has established a scheduling unit with no production commitments. The production is handled by two units, one called *Facts* containing the news department, current affairs, culture, nature and science, and another called *Fiction*, subdivided into music, theatre, entertainment and children's programs.

TV 2 has more or less continued along the same lines as the old organization with its central staff and program planning in one administrative unit, subdivided into a program secretariat, program and technical

Table 17-3. Program Profile of Swedish Television.

	Percent	Hrs/Week*
Educational	8.9	8.2
Religion and ethics	1.3	1.2
News	9.6	8.8
Facts (current affairs and the like)	17.9	16.4
Children's and youth's programs	10.8	9.9
Magazines	5.6	5.1
Sports	9.5	8.7
Music	4.1	3.8
Drama and film	13.6	12.5
Entertainment	13.1	12.0
Miscellaneous	5.6	5.1
	100.0	91.7

* Reruns and regional programs are included, bringing
the total weekly hours up to 91.7, instead of 80 hours, which
is the base output.

planning, economy and personnel, external and procurement. Production is in the hands of five separate units: theatre, entertainment, children's programs, facts, and news.

Technical operations are not handled by the TV channels themselves, but by the TV company. The intention with this is to take advantage of large-scale economy. With 1000 technicians on the payroll in Stockholm alone, and studios and equipment used by both channels, there is ground for capital savings, according to some. Others say that the competitive spirit is lost when more than half of the TV production staff do not belong to any identifiable production department.

The new organization was not wanted by the former Swedish Broadcasting Corporation, but was forced on the company by the government and parliament. The background was partly, as in the rest of Western Europe, a feeling of discontent with media from 1968 and on. This discontent is difficult to describe, the Vietnam War had almost the same impact in Sweden as it had in the United States. The War Resistance Movement engaged in the end almost 100,000 citizens. Even in peaceful Sweden there were some minor riots (no looting, though) and the media did not always report with the detachment and "good taste" responsible politicians wanted.

A wave of "new criticism" swept through media. Swedish press laws are part of the Constitution and thus much more difficult to alter, but the radio laws are easier to change. During the 1960s Swedish Broadcasting Corporation expanded heavily, and even more so during the "golden years" at the beginning of the '70s: the four years from 1968 to 1972 saw an expansion from 3300 to 4300 employees, and at the same time TV personnel doubled. No wonder politicians found it necessary to step on the brake. The main argument was that Swedish Broadcasting Corporation was a heavy, bureaucratic monolith with too many administrators administrating too few program makers. The solution was to cut up the old corporation into five administrative units keeping the three decision-making areas apart: economic planning and control, scheduling power and program making. It is still too early to say if the intended solution will work successfully.

Instead of decreasing the administrative body, the cutting up of the old corporation has led to some doubling of functions. Instead of one board of governors there are now five; instead of one managing director and one deputy managing director, there are five managing directors (but no deputies). It is especially in the regional organization that the anomalities are to be found; the Regional Radio Company has essentially fully staffed regional stations, but due to the new administrative order the TV company and the Radio company now both have their separate regional offices. Instead of 13 regional responsible officers, there are now more than 50.

On the other hand, the new organization gives each company a better chance to supervise the traffic of programs. It might be that in the end, the Swedish Broadcasting Corporation with its manifoldness and its decentralization in a nationally homogenous society will become the effective, slim and popular radio and TV the politicians want it to be.

Integrity, quality, manifoldness and decentralization were the key words the minister of education used when he presented the radio bill to parliament. All the five companies have expressed their goodwill to carry out these promises. The agreement between the five companies on the one hand and the government on the other is valid for seven years, starting 1 July 1979, a considerably short period, but enough, though, for the new organization to settle down. And both media in Sweden are used to changes, as there has been a Royal Commission pondering about how to give the country a decent radio (and in later years, television) in every decade since the 1920s. It is perfectly safe to predict that the 1980s will see a new Royal Commission committed to the same subject.

Chapter 18
Australian Broadcasting

Peter Lucas
Mr. Lucas, a senior officer of the
ABC, has contributed this chapter
in his private capacity; the views
expressed are his own.

For Australians, radio and television represents a backdrop to daily life as universal as the rising of the sun across the island continent they inhabit. The nation's radio services reach virtually every part of Australia's 7,686,884 square kilometers and all but two or three percent of the population is within range of color television.

The choice of program material on offer is as diverse as, on the one hand, a recorded performance by the Antiqua Musica Orchestra of Pertini's Harp Concerto No. 4 in E Flat to, on the other hand, yet another episode of *Days of Our Lives*. While the vast majority of the programs are, not surprisingly, in English, services do, for a variety of purposes, include French, Vietnamese, Neo-Melanesian, Assyrian, Estonian, Russian, Punjabi, Dutch and Yiddish. The quality of sound varies from the purity of stereophonic frequency modulation, available in the captials of the south and eastern seaboard, to the basic message standard of received shortwave in the more remote corners of Australia.

Until the advent of broadcasting, Australians suffered the discomfort (or, depending upon the point of view, the sheltered luxury) of living in the most remote continent in the world, bar Antarctica. Mail from the United States, Britain or Europe took up to a month or more to reach its destination and, the telegraph service excepted, Australia was conditioned to an isolation which exercised a profound effect upon its development. Today, by reaching for a radio or television switch, Australians follow, at first hand, the events which shape our times. Like people in most other Western nations, Australians now devote much of their free time to television viewing and, to a lesser extent, radio listening. On average they spend at least 30 hours a week per household watching television and 22 hours a week per individual listening to radio.

DEVELOPMENT OF RADIO BROADCASTING

While radio began officially in Australia at eight o'clock on the evening of 23 November, 1923, when station 2SB—the first in Australia to operate systematically—transmitted a concert from a Sydney studio, the story of broadcasting in Australia predates that event by some years. The federal government established control over wireless telegraphy in Australia through the Wireless Telegraphy Act of 1905 and, in 1919, assumed power to control wireless telephony. When, in July 1922, Amalgamated Wireless (Australasia) Ltd (AWA) wished to develop broadcasting as a commercial enterprise, it applied, as a matter of course, to the commonwealth for permission to do so.

Smaller firms and amateur experimenters with similar interests objected to the possibility of a monopoly taking over this field and the postmaster-general called a conference of all interested parties. As a result a "sealed set" system of broadcasting was introduced in 1923. By this device, radio sets were sealed to respond to the wavelength of one station only and the licensee paid a listening fee direct to the station. It is a sad yet predictable comment on human nature that the system survived but eight months because of the ease with which listeners could open their sets to receive stations other than those for which they had paid their fees.

In July 1924, the postmaster general issued new regulations which changed the system of licensing and divided radio stations into two categories: A class stations to be financed by listeners' license fees and B class station to be financed by advertising. Under this system, radio broadcasting flourished in the more densely populated areas but proved unprofitable in Tasmania and Western Australia. Country areas were neglected.

Community dissatisfaction with this system crystallized politically in 1928, with action by the federal government to found a national broadcasting service. As licenses of the A class stations expired, studios and technical facilities were acquired and operated by the postmaster general's department; the programs were provided by a commercial contractor, the Australian Broadcasting Company. This arrangement, too, proved to be short-lived.

The following year, an Australian Labor Party Government was elected to power and, in 1931, gave the company notice of termination of its contract. A broadcasting bill was prepared but the government was defeated before the bill could be introduced to parliament. Despite this, the incoming government (representing a different political force, the United Australia Party) adopted the Australian Broadcasting Commission Bill almost unchanged, with the result that the Australian Broadcasting Commission (ABC) was established as from 1 July, 1932. (Under the Australian Broadcasting Act passed in May, 1932; the act has been variously amended over the years and is now consolidated to cover broadcasting in general under the title, the Broadcasting and Television Act 1942).

The B class stations (the term A and B class have long fallen into disuse) continued to exist, thus providing Australia with the basis of the system which, broadly speaking, persists to this day. On the one hand, the ABC provides national and regional services funded from the public purse and carrying no advertising; on the other, the commercial sector operates

314

principally on a local station basis on revenue secured from selling time to advertisers and to sponsors.

F.requency-modulation broadcasting first began in Australia on an experimental basis in 1947 and continued until 1961. Early interest culminated in an inquiry in 1953, which, among other things, focused much upon the technical problems likely to arise with the allocation of VHF channels for both FM radio and television stations. Subsequent hearings (1970/71 and 1974) led to the establishment of an FM service by the ABC in January, 1976, in Sydney, Melbourne, Adelaide and Canberra. The Australian Government in 1979 consented to an extension of FM facilities to commercial radio operators in major capital cities and great interest has been shown in acquiring FM licenses.

DEVELOPMENT OF TELEVISION

In 1953, legislation paved the way for the introduction of television. The following year, following the recommendations of a royal commission, the government approved the ABC as the authority to provide national television programs.

In the commercial sector, licenses were granted following public hearings from interested applicants and television in monochrome commenced in 1956 in Sydney and Melbourne with other capital cities and regional centers following progressively over the next decade or so.

The approach to color television was marked by caution. Many years of engineering, program and administrative planning preceded its launching. Three transmission systems were analysed: (1) the NTSC, developed in 1954 in the United States and now in use in North America, Japan and elsewhere; (2) PAL (Phase Alternation Line) developed in West Germany and used in Britain and Western European countries other than France; and (3) the SECAM system, developed in France and now used there and in the Soviet Union.

After investigation of the merits of these systems and their application to Australian conditions, the PAL system was chosen. The conversion to color involved a complete reorganization of all visual equipment in both ABC and commercial stations—equipment which often dated back to the start of the television monochrome service in 1956.

Australians took to color with remarkable enthusiasm. It took Australia 22 months to achieve the same proportion of color penetration reached in the United States after 10 years of color transmissions. Even at the date of official introduction of color in Australia, 1 March 1975, receivers were already installed in 3% of homes. By November 1979, four out of five homes had color television and the ratio continues to rise. Quite soon monochrome receivers will be as anachronistic as thirsty motor cars.

PUBLIC BROADCASTING AND THE SBS

The 1970s saw two further elements added to the broadcasting spectrum. Public broadcasting stations, financed mainly by community organizations and universities, were established in 1974. Then, in 1978, the Special Broadcasting Service (SBS) was inaugurated to provide, in the words of the enabling legislation, "multilingual broadcasting services and, if authorized by the regulations, to provide multilingual television services . . . and broadcasting and television services for such special purposes as are prescribed."

At this stage, the SBS is responsible for multilingual broadcasting services and has two radio stations operating daily. SBS has also produced foreign-language television programs screened by the ABC for three hours a week.

BROADCAST REGULATION

The regulation of radio and television services in Australia is covered by a variety of acts of parliament, notably the Broadcasting and Television Act 1942; the Parliamentary Proceedings Act 1946; and the Wireless Telegraphy Act. These acts are administered by a federal government minister, the minister for post and telecommunications. (From 1932-1972 the responsibility was carried by the postmaster general, and from 1972-1975 by the minister for the media.)

Under the Australian Constitution the power to make laws with respect to radio and television rests with the federal government, not the individual states. This legislation does not remove the necessity for broadcasters to observe state laws where relevant. The law relating to defamation is an example: national broadcasters have to contend with seven sets of defamation laws.

Australian Broadcasting Commission

The Australian Broadcasting Commission (ABC) provides a national service covering the whole of Australia and an international shortwave service, Radio Australia (Fig. 18-1). Arising from its role as a producer of programs, the ABC also maintains six symphony orchestras, a training orchestra, a show band, and a concert organization (Figs. 18-2 and 18-3). It is a book and magazine publisher on a considerable scale too.

The governing body of the ABC consists of a commission of between nine and 11 members, at least two of whom must be women. By law, each of the six Australian states must be represented on the commission by a nominated commissioner. Commissioners are appointed from three to five years and their appointments may be renewed.

Responsibility for the ABC's program output is entrusted to the commission, with freedom from control by the Government of the day. The commission has absolute power to determine what the ABC shall carry by way of programs, including those of a controversial or political nature. It is true that, under the Broadcasting and Television Act, the minister to whom the ABC reports can require the ABC to broadcast or televise a particular item which he considers to be in the national interest; he can also require the commission to refrain from broadcasting a particular item. He has similar powers in relation to Australian commercial stations. If the minister chooses to exercise this power, the ABC and

Fig. 18-1. Broadcast House, the administrative headquarters of the Australian Broadcasting Commission, Sydney.

commercial stations are equipped with safeguards. The minister is required to report the circumstances to parliament within seven sitting days. In addition the ABC is required to report, in its annual report to parliament, any use of this power by the minister in respect of ABC programs. In practice, the use of these powers is avoided by ministers, so far as the ABC is concerned only four times in almost 50 years, the last being in 1963.

One of the commission's primary responsibilities is set out in the Parliamentary Proceedings Broadcasting Act which provides for the radio broadcasting of proceedings in the federal parliament. This obligation consumes some eight hours of air time on one of the ABC's three radio networks on each day parliament is sitting, about 76 days a year. There is also provision for the televising of proceedings of parliament. But, except for telecasts of the ceremonial opening at the beginning of each new session of parliament every two or three years, the only occasion on which parliamentary proceedings have been televised was in 1974 when the ABC televised a joint sitting of both houses (the house of representatives and the senate).

The ABC is not permitted to broadcast advertisements on its radio and television services. It operates principally on funds provided by federal parliament. At the end of 1979, the commission employed a full-time staff of just over 6000 people and its appropriation for the 1978/79 financial year was $Aust146 million. A further $Aust12million was obtained from receipts derived mainly from such activities as public concerts and merchandising of sound cassettes of programs, books, booklets and other program-related products.

The ABC has radio and television stations in the Australian national capital of Canberra, in each of the six State capitals and in Darwin, the capital of the Northern Territory, which is due for full statehood in the near future. In addition, radio transmitters are located in or around all country towns and cities. Thirty regional centers have radio studios and are equipped for local production; three regional centers have studio facilities for television.

ABC radio services are transmitted for up to 24 hours a day through a total of 89 medium-wave and six shortwave stations. The ABC's overseas

Fig. 18-2. Louis Fremaux; chief conductor Sydney Symphony Orchestra. The SSO is one of seven ABC orchestras.

318

Fig. 18-3. Brian May and the ABC Melbourne Showband. This band is widely used for middle-of-the road recordings, radio broadcasts and ABC television programs.

service, Radio Australia, is carried by 13 shortwave transmitters and is broadcast around the clock. Radio Australia is on the air in nine languages: English, French, Thai, Japanese, Indonesian, Standard Chinese, Cantonese, Vietnamese and Neo-Melanesian. While its broadcasts can be heard in all continents, particular emphasis programming is paid to South and Southeast Asia.

ABC television programs are transmitted on an average of 12 hours a day and are available through 85 transmitters and 80 translator stations.

Transmitters for ABC services are provided by the Australian Telecommunications Commmission under arrangements approved by the minister for post and telecommunications.

The ABC has overseas offices in London (since 1932), New York (1942), Washington, Brussels, Singapore, Kuala Lumpur, Djakarta, Tokyo, New Delhi, Peking, Bangkok and Port Moresby.

In association with civic and state government authorities, the ABC maintains symphony orchestras in each of the six states of Australia. The orchestras' permanent combined strength totals about 400 players in addition to which the ABC supports a training orchestra of graduate students. The ABC is, in fact, one of the largest concert entrepreneurs in the world and each year presents about 750 public concerts.

In news, the ABC is distinctive in that within Australia, through word, film and videotape, it is the nation's most comprehensive newsgatherer. The ABC is obliged under the Broadcasting and Television Act to collect news from within Australia with its own staff. The service employs some 300 journalists who are stationed in leading cities and regional centers and calls on the services of about 2000 part-time correspondents scattered across the country. To supplement overseas agency services, the ABC also maintains staff correspondents in North America, Britain, Europe and Asia.

Bound up with its role in exploring the art of radio and television in Australia, the commission has contributed to the development of public taste in music, drama and in other fields as well as operating specialized services such as educational broadcasting and a national rural service. It entertains, and provides the facts and background in which informed public opinion may flourish. Against this background, the ABC continues to exercise a major influence on the quality of life in Australia.

Australian Broadcasting Tribunal

Commercial and public broadcasters are licensed individually by a government regulatory authority, and the Australian Broadcasting Tribunal. The tribunal came into existence on 1 January, 1977, replacing the Australian Broadcasting Control Board. As well as its licensing function, the tribunal also determines standards to be observed by licensees and oversees the suitability of program material. The tribunal is empowered to hold public inquiries on these and other related matters, including investigations into breaches of licensing conditions. When dealing with the granting, renewing, suspension or relocating of licenses, the tribunal may impose conditions or penalties.

Since its establishment, the tribunal has conducted a survey about the wishes of the community regarding broadcasting services. The terms of reference of this inquiry include whether control of advertising standards should be left to the broadcasters, the extent of Australian content in programs and general program matters such as the nature of children's programs and political and religious matter in broadcasting. The tribunal also conducts public hearings about the renewal of commercial television licenses, with particular emphasis on the content and nature of children's programs. The tribunal's wide powers indicate that future inquiries and hearings will deal with other aspects of broadcasting in Australia.

The ABC is outside the responsibility of the Australian Broadcasting Tribunal.

Commercial Services

There are 125 commercial radio stations, primarily local stations, broadcasting to local communities. Although more than half are member stations of one or another of network groups, national networking, in the sense that it applies to the ABC, does not exist. Some groups do have arrangements for relaying certain programs such as sport or news. There are 50 commercial television stations.

Commercial radio and television stations are privately owned and rely for their income almost entirely upon the transmission of advertisements. As the Federation of Australian Commercial Television Stations has pointed out: "Advertising is not an optional extra, not a separate element, not a disposable luxury. It is an integral and inseparable component of the system, of its operation, of its livelihood, as important as its programming."

The commercial sector applies its resources vigorously to the pursuit of attracting revenue—and with great success. In a population of 14.2 million people, 13.6 million of the population are estimated to live in homes with television receivers; radio exists virtually in 100% of homes.

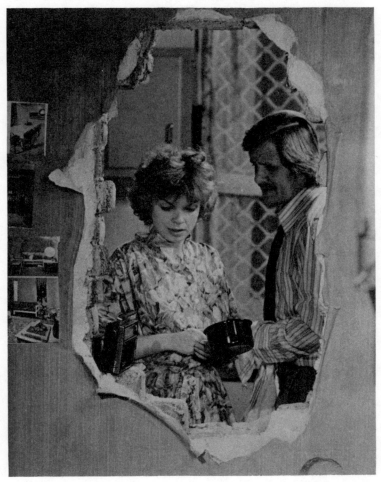

Fig. 18-4. *Tickled Pink–One Day Miller* is an ABC situation comedy. Each play is self-contained. Pictured is Miller (Tony Llewelyn Jones) and his wife Marilyn (Penne Hackforth Jones).

Advertising expenditure on radio and television totalled $Aust403 million in 1977—more than twice the amount spent on the ABC.

With commendable frankness, commercial representatives point out: "Because the nature of commercial broadcasting is such that it requires and is dependent upon the availability of large audiences, we have become very skilful at providing programming of the widest popular appeal. (See Figs. 18-4, 18-5, 18-6, and 18-7.) In turn, this means that the commercial sector is not well-suited to catering for the needs of minority or specialty interest groups whose requirements, by definition,are far more highly specialized than the mass audience. The competitive nature of commercial broadcasting means that it is highly responsive to changes in audience preference, since advertising revenue is also highly volatile. Thus, the commercial sector is very adept at catering for mass audiences." (Submission by the Federation of Australian Commercial Television Stations to the Committee of Inquiry Into the Broadcasting Industry, June 1976.)

Special Broadcasting Service

The Special Broadcasting Service (SBS) had its genesis in 1975 when two experimental stations were set up to provide programs for ethnic groups. They were in Sydney and Melbourne and were called 2EA and 3EA respectively; EA meaning Ethnic Australia. When it is realized that there are more Maltese in Australia than in Malta, and that Melbourne is said to be the city with the third largest Greek-speaking population in the world, some idea of the potential audience for ethnic broadcasts can be gained.

There were seven language groups in the original experimental service in Sydney: Italian, Greek, Turkish, Spanish, Arabic, Maltese and four language transmissions to serve the Yugoslav community. The Melbourne station developed along separate but similar lines.

The second stage of 2EA's development saw the introduction of Russian, Mandarin, Cantonese, German and Polish. Since then further languages have been added including Latvian, Dutch, Ukranian, Portugese, French, Vietnamese, Byelorussian, Assyrian, Slovak, Lithuanian, Hungarian, Armenian, Estonian, Hebrew, Yiddish, and seven languages of the Indian subcontinent: Hindi, Urdu, Bengali, Tamil, Punjabi, Marathi and Gujarati.

This experiment may be unique in domestic broadcasting services. There are multilingual stations in other countries, but perhaps none which broadcast in so many languages—37 at the present time, with two others, Tagalog and Rumanian to be added shortly. There are many philosophical considerations about broadcasting special programs for so many ethnic groups whose numbers, even collectively, are still a minority of the Australiam population, and there is continuing debate about what such a service should do.

The Special Broadcasting Service was established in 1978 by amendment to the Broadcasting and Television Act. Like the ABC, the

Fig. 18-5. *On Jazz* is a half-hour ABC jazz series featuring the Ken Herron Jazz Band and special guests recorded before an audience at the Melbourne Hotel, Brisbane. Pictured (left to right) Tony Ashby, Bob Watson, Bob Barnard, Horsley Dawson, Ken Herron and Graham Bell.

SBS is a statutory authority funded by an appropriation from federal parliament.

Public Broadcasting

For many years, radio broadcasting in Australia had been restricted to no more than six commercial radio and two ABC stations in each capital city and one commercial and one ABC station to serve the main provincial centers.

In 1973, the body then administering frequency allocation, the Australian Broadcasting Control Board, re-examined the question of frequency allocation and established that a public broadcasting system was technically feasible. The government of the day decided that there was a social need for the introduction of broadcasting services to accommodate those community needs beyond those being provided by the existing ABC and commercial radio stations. Such stations were to develop from "a significant and spontaneous community demand" and to be community-funded in most cases. Funds could come from a limited form of program sponsorship and from subscription; or from the support of such bodies as universities and colleges of advanced education.

The first station to go on air—the Adelaide University-based VL-5UV (now 5UV)—began broadcasting in 1972, initially to the university campus and subsequently (1974) to the public. On 15 December, 1974, the first community subscription-funded station began broadcasting, 2MBS-FM, Sydney.

There are now 25 public broadcasting stations licensed under the Broadcasting and Television Act. Of these, 20 are on air; the remainder are still setting up studios and facilities. Public broadcasting stations are required to operate on a nonprofit basis and to offer programming distinctly different from the ABC or the commercial stations. Advertising of the spot announcement type is not permitted in public broadcasting. Sponsorship has to be limited to a form approved by the minister and a watch is kept by the Australian Broadcasting Tribunal on the extent of sponsorship.

THE FUTURE

Predicting the future of broadcasting is a somewhat unprofitable occupation. There is perhaps only one certainty, you will never get it totally right. What does emerge as a continuing theme, however, is that Australians agree strongly with the concept of a publicly-funded radio and television service. In terms of audience patronage the ABC secures consistent support. In Melbourne, for example, 80% of homes tune to ABC television at some time during the course of a week. In smaller towns and rural areas the figure can run higher.

Fig. 18-6. Scene from the ABC TV production *The Inventors*. Shown is a prototype solar tracker. ABC conducts an Inventor of the Year Award and bases a series of feature programs on Australian inventions. The ABC produces about 60% of its own programs on television.

Fig. 18-7. *The Truckies.* Shown are the two leads in the series, John Wood and Michael Aitkens, who played Stokey and Chris.

As an organization free of commercial pressures as to its programming, the ABC is required to provide radio and television services which are, to quote the Broadcasting and Television Act, "adequate and comprehensive." The ABC does not have to chase ratings. Indeed its governing board, the commission, has actively criticized the attention devoted by the Australian media to such criteria, arguing that ratings reveal nothing about the quality of programming nor about the attention or interest generated by programs. The community accepts the ABC as an integral part of the fabric of Australian society. A public opinion poll conducted in August/September 1979 by the research organization,

325

Hoare, Wheeler and Lenehan, found that "nine persons in 10 (87%) acknowledge that the country benefits from having a national broadcasting system. And eight in 10 (78%) support the use of public monies in running the services."

So far as the commercial sector is concerned there is not the slightest doubt of the appeal it makes to the mass market. The usual viewing and listening pattern is that, in terms of audience "tonnage," commercial radio and television programs draw much the bigger crowd. This is partly traceable to sympathetic marketing backup. Most radio and television stations are linked to newspaper and magazine groups and can expect to receive valuable in-house publicity for their programs. More importantly, perhaps, is the fact that commercial radio and television unashamedly— and for the best of reasons—sets out to maximize its audiences. The reasons are quite clear cut. Advertising is the lifeblood of commercial radio and television and advertising rates are geared to the size of audiences delivered. Thus commercial radio and television's continuing theme is to seek out optimum popularity. The formula works. The Study of Listener Attitudes (February, 1979), conducted by Audience Studies, Incorporated Australia Pty. Ltd., for the Federation of Australian Radio Broadcasters found that the commercial radio listening market was characterised by "a high degree of satisfaction with the product and service it is receiving." The study continued: "High levels of satisfaction are accompanied by a concomitant lack of interest in issues other than immediate programming issues.

"Awarenness of systemic issues, social, political or philosophical considerations affecting the broadcasting system in this country is very low.

"Radio is seen as an essential, habitual, instinctively consumed product. Its advantages are identified as its mobility, its immediacy and its undemanding nature. It is viewed as more necessary but less involving than television or press."

Advertisers, driven by a regard for ratings, have no qualms about investing heavily in air time on commercial stations. In 1977 $Aust403 million was spent on advertising on radio and television, and it can be anticipated that subsequent figures, when published will run higher. In the case of commercial television the expenditure in 1977—$Aust304 million—was triple the figure for six years earlier.

A submission by the Federation of Australian Commercial Television Stations (February 1977) to the Australian Broadcasting Tribunal carried the observation: "Soaring program and production costs, especially of local material, have joined with this heavy demand to produce an inevitable increase in advertising rates. Nevertheless, commercial television is still more than competitive with other media in terms of cost efficiency. Advertisers have realized the fact. In five years, television has picked up a steadily increasing share of total advertising expenditure."

It would be foolhardy indeed to suggest that political and public attitudes will remain fixed as Australia moves through the '80s and '90s. Undoubtedly new hardware will affect the scenario: the introduction of a domestic satellite system (announced by the federal government, October 1979) and the marketing of cheap home vidoecassette facilities are two examples. Yet other sweeping developments have taken place in the past: the upgrading of AM signal quality, FM, the advent of television, first in black and white, then in color. And none of this much affected the broad support by Australians for the mixture of the public/private broadcasting system.

As a people, Australians are not inclined to quarrel with the radio and television system they have established. In itself that is a tribute to the stability of a basic approach now tested across half a century or more.

SOURCE REFERENCES

Submission to Standing Committee on Education, Science and the Arts - Senate of the Australian Parliament, 6 June, 1972.

Australian Broadcasting, submission by the Federation of Australian Commercial Television Stations to the Committee of Inquiry into the Broadcasting Industry (June, 1976).

Australian Broadcasting, Report on the structure of the Australian broadcasting system (F.J. Green, Secretary of the Postal and Telecommunications Department). September, 1976.

A Critical Appraisal, by the Federation of Australian Radio Broadcasters. (December 1976).

Submission to the Australian Broadcasting Tribunal by the Federation of Australian Commercial Television Stations 1977/78.

An Ethnic Television Service for Australia - Discussion paper by the Ethnic Television Review Panel 1978.

Australian Commercial Radio - A Study of Listener Attitudes - Federation of Australian Radio Broadcasters (February, 1979)

Radio 1979 (Federation of Australian Radio Broadcasters)

TV Facts 4 (Federation of Australian Commercial Television Stations 1978)

History and Development of ABC (1979)

Your ABC (1979)

ABC 43rd/47th Annual Reports to Parliament

Status of the Media 1979 (George Patterson Pty. Ltd. Sydney)

Appendix:
Additional Reading

GENERAL

Abshire, David. *International Broadcasting: A New Dimension of Western Dipolmacy*. Beverly Hills; Sage, 1976.

Adhikarya, Ronny. *Broadcasting in Peninsular Malaysia*. London; Routledge and Kegan Paul, 1977.

Arnove, Robert F., Ed. *Educational Television: A Policy Critique and Guide for Developing Countries*. New York: Praeger, 1976.

Briggs, Asa. *The History of Broadcasting in the United Kingdom*, Vol. I: *The Birth of Broadcasting*, 1961, Vol. II: *The Golden Age of Wireless*, 1965, Vol. III: *The War of Words*, 1970, London: Oxford University Press.

Carter, Martin D. *An Introduction to Mass Communications: Problems in Press and Broadcasting*. New York: Humanities Press, 1971.

Cherry, Colin. *World Communication: Threat or Promise?* New York: John Wiley, 1978.

Dowling, Jack; Doolan, Lelia and Quinn, Bob. *Sit Down and Be Counted: The Cultural Evolution of a Television Station*. Dublin; Wellington, 1969.

Eguchi, H. & H. Ichinotte, Eds. *International Studies of Broadcasting with Special Reference to the Japanese Studies*. Tokyo: NHK Radio and TV Culture Research Institute, 1971.

Emery, Walter B. *National and International Systems of Broadcasting: Their History, Operation, and Control*. East Lansing: Michigan State University Press, 1969.

Fischer, Heinz-Dietrich & John E. Merrill, Eds. *International and Intercultural Communication*. New York; Hastings House Publishers, 1976.

Gerbner, George, Editor. *Mass Media Policies in Changing Cultures*. New York: John Wiley & Sons, 1977.

Goodhart, G.J; A.S.C. Ehrenberg; & M.A. Collins. *The Television Audience: Patterns of Viewing*. Lexington, Mass: Lexington Books, 1975.

Gorman, Maurice. *Broadcasting & Television Since 1900*. London; Andrew Dakers Ltd., 1952.

Green, Timothy. *The Universal Eye, the World of Television*. New York: Stein & Day, 1972.

Hallman, Eugene S. & H. Hindley. *Broadcasting in Canada*. Don Mills, Ontario: General Publishing Co. Ltd., 1977.

Harasymiw, Bohdan, Ed. *Education and the Mass Media in the Soviet Union and Eastern Europe*. New York: Praeger, 1976.

Harris, Paul. *When Pirates Ruled the Waves*. Blue Ridge Summit, Pa: TAB Books, 1972.

Head, Sydney W., Ed. *Broadcasting in Africa: A Continental Survey of Radio and Television*. Philadelphia: Temple University Press, 1976.

Hollander, Gayle Durham. *Soviet Political Indoctrination: Developments in Mass Media and Propaganda Since Stalin*. New York: Praeger, 1972.

Hopkins, Mark W. *Mass Media in the Soviet Union*. New York: Pegasus, 1970.

Katz, Elihu & George Wedell. *Broadcasting in the Third World*. Cambridge: Harvard University Press, 1978.

Kyokai, Nippon Hoso. *The History of Broadcasting in Japan*. Tokyo: Radio and Television Culture Research Institute, 1967.

Lerner, Daniel & Wilbur Schramm, Ed. *Communication and Change in Developing Countries*. Honolulu: East-West Center Press, 1967.

Levoi, Bela, *Hungarian Radio and Television 1970*. Budapest: Hungarian Radio & TV, 1970.

Lichty, Lawrence W. *World & International Broadcasting: A Bibliography*. Washington D.C.: Association for Professional Broadcasting Education, 1971.

Lisann, Maury. *Broadcasting in the Soviet Union: International Politics and Radio*. New York, Praeger, 1975.

McWhinney, Edward, Ed. *The International Law of Communications*. Leyden, Netherlands: A.W. Sijthoff, 1971.

Paulu, Burton. *Radio & Television Broadcasting in Eastern Europe*. Minneapolis: University of Minnesota Press, 1974.

Paulu, Burton. *Radio & Television Broadcasting on the European Continent*. Minneapolis: University of Minnesota, 1967.

Polman, Edward W *Broadcasting in Sweden*. Boston: Routledge & Kegan Paul, 1974.

Public Opinion and Mass Communication. Budapest: Mass Communication Research Center, Hungarian Radio and Television, 1972.

Quicke, Andrew. *Tomorrow's Television: An Examination of British Broadcasting Past, Present, and Future*. Berkhamsted, England: Lion Publishing, 1976.

Reynolds, Michael M. *A Guide to Thesis' & Dissertations: An Annotated International Bibliography of Bibliographies*. Detroit, Michigan: Gale Research Company, 1975.

Sherman, Charles & Donald Browne, Eds. *Broadcasting Monographs No. 2: Issues in International Broadcasting*. Washington, D.A.: Broadcast Education Association, 1976.

Wells, Alan. *Picture Tube Imperialism? The Impact of U.S. Television on Latin America*. Maryknoll, N.Y.: Orbis Books, 1972.

Wilcox, Dennis L. *Mass Media in Black Africa: Philosophy & Control*. New York: Praeger, 1975.

World Radio TV Handbook. Hvidoure, Denmark. Annual.

SOUTH AFRICA

Erasmus P.F. *Die radio as massa-kommunikasiemedium met spesiale verwysing na die situasie in Suid-Afrika*. Human Sciences Research Council—Report Komm I, Pretoria 1970.

Orlik P. *The South African Broadcasting Corporation—An Historical Survey and Contemporary Analysis*. University Microfilms, Ann Arbor, Michigan, 1968.

Strydom G.S. *Die totstandkoming en uitbouing van die skoolradiodiens van Radio Bantu*. Human Sciences Research Council Report No 0-55, Pretoria, 1975.

South African Broadcasting Corporation, *Annual Report*, Johannesburg, 1938.

South African Broadcasting Corporation, *Annual Report*, Johannesburg, 1950.

South African Broadcasting Corporation, *Annual Report*, Johannesburg, 1954.

South African Broadcasting Corporation, *Annual Report*, Johannesburg, 1961.

South African Broadcasting Corporation, *Annual Report*, Johannesburg, 1978.

Van Vuuren D.P., Puth G. and Roos D. *The scope and impact of the health guidance programme, "Impilo Yethu" of the Xhosa Service of the SABC*. HSRC Report Comm N-3, Pretoria, 1977.

South Africa '77, Official Yearbook of the Republic of South Africa, 1977, Government Printer, Pretoria, 1978.

CANADA

Allard, T.J. *The C.A.B. Story 1926-1976: Private Broadcasting in Canada*. Ottawa: Canadian Association of Broadcasters, 1976.

Alyluia, Kenneth Joseph. *Regulation of Commercial Advertising in Canada*. rev. ed. Winnipeg: University of Winnipeg, 1972.

Babe, Robert E. *Cable Television and Telecommunications in Canada*. East Lansing, Michigan: Michigan State University, 1974.

Beswick, Peter A. *The Board of Broadcast Governors and the Public Interest*. Thesis, School of Public Administration, Carleton University, Ottawa, 1962.

Buttrum, Keith. *The Cable TV Industry in Canada*. Basic Issues in Canadian Mass Communication. Montreal, McGill University, 1973.

Canada. Department of Communications. *Communications: Some Federal Proposals*. Ottawa: Information Canada, 1975.

Canada. Ministre des communications. *Telecommunications: quelques propositions federales*. Ottawa: Information Canada, 1975.
    Proposals for a Communication Policy for Canada: A Position Paper of the Government of Canada. Ottawa: DOC, 1973 (Green paper).

Canada. Ministere des communications. *Vers une politique nationale des telecommunications: expose du governement du Canada*. Ottawa: MDC, 1973 (Livre vert)

Canada. Department of Communications. Telecommission Directing Committee. *Instant World: A Report on Telecommunications in Canada*. Ottawa: Information Canada, 1971.

Comite directeur de la telecommission. *Univers sans distance: rapport sur les telecommunications au Canada*. Ottawa: Information Canada, 1971.

Department of Industry. *White Paper on a Domestic Satellite Communication System for Canada*. Ottawa: Queen's Printer, 1968.

Ministere de l'industrie. *Livre blanc sur un systeme de telecommunications pour le Canada*. Ottawa: Imprimeur de la Reine, 1968.

Department of the Secretary of State. *White Paper on Broadcasting, 1966*. Ottawa: Queen's Printer, 1966.

Ministere du Secretariat d'Etat. *Livre blanc sur la radiodiffusion*, 1966. Imprimeur de la Reine, 1966.

Laws, Statutes, etc. *Broadcasting Act, 1967-68, c. 25, s. 1*. Ottawa: Queen's Printer, 1970.

Lois, statuts, etc. *Loi sur la radiodiffusion, 1967-68, c. 25, art. 1*. Ottawa: Imprimeur de la Reine, 1970.

Canadian Radio-Television and Telecommunications Commission Act. Ottawa: Queen's Printer, 1975.

Loi sur le Conseil de la radiodiffusion et des telecommunications canadiennes. Ottawa: Imprimeur de la Reine, 1975.

Parliament. House of Commons. *Minutes of Proceedings and Evidence of the Standing Committee on Broadcasting, Films and Assistance to the Arts*. Ottawa: Queen's Printer, 1966.

Parlement. Chambre des Communes. *Proces-verbaux et temoignages du Comite permanent de la radiodiffusion, des films et de l'assistance aux arts*. Ottawa: Imprimeur de la Reine, 1966.

Senate. Special Committee on Mass Media. *Report*. Ottawa: Queen's Printer, 1970 (Davey Committee).

Senate. Comite special sur les moyens de communications de masse. *Rapport*. Ottawa: Imprimeur de la Reine, 1970 (Rapport Davey).

Royal Commission on Broadcasting. *Report*. Ottawa: Queen's Printer, 1957, (Fowler Report).

Commission royale d'enquete sur la radio et la television. *Rapport* Ottawa, Imprimeur de la Reine, 1957 (Rapport Fowler).

331

Royal Commission on National Development in the Arts Letters and Sciences. *Report*. Ottawa: King's Printer, 1951, (Massey Commission).

Commission royale d'enquete sur l'avancement des arts, lettres et science au Canada. *Rapport*. Ottawa: Imprimeur du Roi, 1951 (Commission Massey).

Royal Commission on Radio Broadcasting. *Report*. Ottawa: King's Printer, 1929 (Aird Report).

Commission royale de la radiodiffusion. *Rapport*. Ottawa, Imprimeur du Roi, 1929 (Rapport Aird).

Canadian Broadcasting Society. *CBC: A Brief History and Background*. Ottawa: CBC Information Services, 1972.

Societe Radio-Canada. *Petite histoire de la Societe Radio-Canada*. Ottawa: Service de l'information externe, Radio-Canada, 1972.

Canadian Radio-Television and Telecommunications Commission. *Annual Report*. Ottawa: Information Canada, 1968.

Conseil de la radiodiffusion et des telecommunications canadiennes. *Rapport Annuel*. Ottawa, Information Canada, 1968.

"Radio Frequencies are Public Property; Public Announcement and Decision of the Commission on the Applications for Renewal of the Canadian Broadcasting Corporation's Television and Radio Licenses." Ottawa: CRTC, 1974 (CRTC Decision 74-70).

"Les ondes radioelectriques sont propriete publique; avis public et decision du Conseil sur les demandes de renouvellement des licences de television et de radio de la Societe Radio-Canada." Ottawa: C.R.T.C., 1974 (Decision C.R.T.C. 74-70).

A Resource for the Active Community. Ottawa: Information Canada, 1974.

Radiodiffusion et communaute. Ottawa: Information Canada, 1974.

Research Branch. *Directory: Multilingual Broadcasting in Canada*. Ottawa: Information Canada, 1974.

Direction de la recherche. *Annuaire de la radiodiffusion multilingue au Canada*. Ottawa: Information Canada, 1974.

Research Branch. *Multilingual Broadcasting in the 1970s*. Ottawa: Information Canada, 1974.

Direction de la recherche. *La radiodiffusion multilingue dans les annees 70*. Ottawa: Information Canada, 1974.

Committee of Inquiry into the National Broadcasting Service. *Report*. Ottawa: CRTC, 1977.

Comite d'enquete sur le service national de radiodiffusion. *Rapport*. Ottawa: C.R.T.C., 1977.

Cooper, Harriet. *Advertising in Canada*. Basic Issues in Canadian Mass Communication. Montreal: McGill University, 1973.

Regulation and Control in Canadian Mass Media. Basic Issues in Canadian Mass Communication. Montreal: McGill University, 1973.

Cossette, Claude. *Communication de masse et consommation de masse*. Sillery, Quebec: Editions du Boreal Express, 1975.

English, H. Edward. *Telecommunications for Canada: An Interface of Business and Government*. Toronto: Methuen, 1973.

Firestone, O.J. *Broadcast Advertising in Canada: Past and Future Growth*. Ottawa: University of Ottawa Press, 1966.

Grant, Peter S. *Broadcasting and Cable Television Regulatory Handbook: Statutes, Regulations, Procedures, Standards and Other Information Relating to the Regulation of Broadcasting Stations and Cable Television Ssytems in Canada*. Toronto: Law Society of Upper Canada, 1973.

Recueil legislatif sur la radiodiffusion et la television par cable: les statuts, reglements, procedures, normes et d'autres rensengnements qui ont rapport a la reglementation des postes de radiodiffusion et des systemes de television par cable au Canada. Toronto: Law Society of Upper Canada, 1973.

Hallaur, E.S. *Broadcasting in Canada*. Don Mills, Ontario: General Publishing, 1977.

Institut canadien d'education des adultes. *La radiodiffusion au Canada depuis ses origines jusqu'a nos jours*. Cahiers d'information et de documentation, nos 16 et 17. Montreal: I.C.E.A., 1964.

Irving, John A. *Mass Media in Canada*. Toronto: McGraw-Hill Ryerson, 1962.

Jamieson, Donald C. *The Troubled Air: A Frank, Thorough and Sometimes Disturbing Look at the Present State of Canadian Braodcasting by A Noted Broadcaster*. Fredericton, NB: Brunswick Press, 1966.

Lambert, Richard Stanton. *School Broadcasting in Canada*. Toronoto: University Of Toronto Press, 1963.

Linton, James and Edmunds, Hugh. *Canadian Television Viewing Habits: Present Patterns and Future Prospects*. Windsor: Centre for Canadian Communications Studies, 1976.

McKay, Bruce. "The C.B.C. and the Public; Management Decision Making in the English Television Service of the Canadian Broadcasting Corporation, 1970-1974; A Doctoral Dissertation Case Study." Stanford, California: California Institute for Communication Research, 1976.

Malloch, Kati. *Broadcasting in Canada*. Basic Issues in Canadian Mass Communication. Montreal: McGill University, 1973.

Malone, William. *Broadcast Regulation in Canada: A Legislative History*. Thesis, Harvard University, 1962.

Matlin, Evelyn. *The Record Industry in Canada*. Basic Issues in Canadian Mass Communication. Montreal: McGill University, 1973.

Matthew's List. Meaford, Ontario: S. Matthews, 1956 (quarterly).

O'Brien, John Egli. *A History of the Canada Radio League: 1930-1936*. Thesis, University of Southern California, Los Angeles, 1964.

Ontario Royal Commission on Violence in the Communications Industry. *Final Report*. 7 volumes, Vol. 1: *Approaches, Conclusions and Recommendations*. Vol. 2: *Violence and the Media: A Bibliography in Print and Music*. Vol. 5: *Learning from the Media*. Vol. 6: *Vulnerability to Media Effects*. Vol. 7: *The Media Industries: From Here to Where*? Toronto: Queen's Printer, 1977 (LaMarsh Commission Report).

Ontario. Royal Commission on Violence in the Communications Industry. *Rapport final*. Vol. 1: *Des approaches, des conclusions, des recommendations*. Toronto: Imprimeur de la Reine, 1977 (Rapport LaMarsh).

Peers, Frank. *The Politics of Canadian Broadcasting, 1920-1969*. Toronto: University of Toronto Press, 1969.

Pitfield, McKay, Ross & Co. Ltd. *The Canadian Broadcasting Industry: Special Report*. Toronto: Pitfield, McKay, Ross, 1971.

Preshing, William Anthony. "The Canadian Broadcasting Corporation's Commercial Activities and their Interrelationship to the Corporation's Objectives and Development." Thesis, University of Illinois, 1965.

Romanow, Walter Ivan. "The Canadian Content Regulations in Canadian Broadcasting: An Historical and Critical Study." Thesis, Wayne State University, Detroit, 1975.

Rosen, Earl, ed. *Educational Television Canada: The Development and State of ETV, 1966*. Toronto: Burns and MacEachern, 1967.

Saunderson, Hugh Lawrence. "Broadcasting and Regulation: The Growth of the Single System in Canadian Broadcasting." Thesis, Carleton University, Ottawa, 1972.

Shea, Albert. *Broadcasting the Canadian Way*. Montreal: Harvest House, 1963.

Singer, Benjamin, D. *Communications in Canadian Society*. 2nd rev. ed. Toronto: Copp Clark, 1975.

Toogood, Alexander Featherstone. *Canadian Broadcasting: A Problem of Control*. Ottawa: Canadian Association of Broadcasters, 1969.

⸻⸻⸻. *Canadian Broadcasting in Transition: 1919-1969*. Ottawa: Canadian Association of Braodcasters, 1961.

Weir, Ernest Austin. *The Struggle for National Broadcasting in Canada*. Toronto: McClelland and Stewart, 1965.

Index